WHEN MOUNTAIN LIONS ARE NEIGHBORS

The National Wildlife Federation and Heyday gratefully acknowledge the individuals and organizations whose generous contributions made this publication possible:

WILDLIFE HEROES

Jon Christensen

Susan Gottlieb To my husband, Dan, for his amazing generosity and willingness to support Beth's important work.

David Crosby

WILDLIFE LEADERS

Furthermore: A Program of the J. M. Kaplan Fund

Dan and Debbie Gerber

Annemarie Hoffman I have firsthand experience of Beth's love for and commitment to pikas and all the other natural wonders of Yosemite. Any opportunity we have to create healthy respect for, and cohabitation and cooperation with, the animal world is a moment blessed with grace.

Jerald and Madelyn Jackrel To Rebecca, for the beauty she shared with us all.

Charles and Doris Michaels For Marty and Denali Schmidt, who loved nature, the mountains, and wildlife.

Jeremy Railton

Steve and Rosemarie Smallcombe

Jerry Voight and Jean Burke Fordis To Beth, whose commitment to protecting wildlife inspires us each and every day.

WILDLIFE FRIENDS

William Akel, Kathi and George Colman and Family, Faith Hershiser, Susan Mokelke, Gebhard and Liana Neyer, Wayne and Brooke Schick, Melani Smith, Kim Monday, Christine and Stephan Volker, Lynn Wells

Proceeds from the sale of this book directly benefit the NWF's wildlife conservation work in California.

NATIONAL
WILDLIFE
FEDERATION®

www.nwf.org/california

When Mountain Lions Are Neighbors

PEOPLE AND WILDLIFE WORKING IT OUT IN CALIFORNIA

Beth Pratt-Bergstrom

Foreword by Collin O'Mara

Heyday, Berkeley, California
The National Wildlife Federation, Reston, Virginia

Library of Congress Cataloging-in-Publication Data
Names: Pratt-Bergstrom, Beth. | National Wildlife Federation.
Title: When mountain lions are neighbors : people and wildlife working it out
    in California / Beth Pratt-Bergstrom.
Description: Berkeley, California : Heyday ; Reston, Virginia : National
    Wildlife Federation, [2016] | Berkeley, CA : [Distributed by] Heyday
Identifiers: LCCN 2015043694 (print) | LCCN 2015044823 (ebook) | ISBN
    9781597143462 (pbk. : alk. paper) | ISBN 9781597143486 (kindle)
Subjects: LCSH: Wildlife habitat improvement--California. | Wildlife
    conservation--California. | Human-animal relationships--California. |
    Urban animals--California. | Suburban animals--California. | Wildlife
    crossings--California.
Classification: LCC QL84.22.C2 P73 2016 (print) | LCC QL84.22.C2 (ebook) |
    DDC 333.95/409794--dc23
LC record available at http://lccn.loc.gov/2015043694

Front cover photo: Steve Winter. View his work at www.stevewinterphoto.com.
Back cover photos from top: Rebecca Abbey, Karl Frankowski, Steve Winter, Robert E. Riggins, National Park Service
Book design: Rebecca LeGates
Poem on page 41 ©2016 Nick Asbury, used with permission.

Orders, inquiries, and correspondence should be addressed to:
    Heyday
    P.O. Box 9145, Berkeley, CA 94709
    (510) 549-3564, Fax (510) 549-1889
    www.heydaybooks.com

Printed in East Peoria, IL, by Versa Press, Inc.

FSC
www.fsc.org
MIX
Paper from
responsible sources
FSC® C005010

10 9 8 7 6 5 4 3 2 1

FOR P-22

P-22 roaming Griffith Park in 2016.

CONTENTS

"There is a wildness in California still....It does not exist only within those areas designated as wilderness or national parks, but also in the 'back blocks'—the hinterland of lands that are otherwise used for other purposes. There is a feeling that back beyond the next ridge there is still wild country. And even beyond that there is an invasion of the wild into the tame that for some of us brings feelings of security."—Raymond Dasmann, *Wild California: Vanishing Lands, Vanishing Wildlife*

"What is the message that wild animals bring, the message that seems to say everything and nothing? What is this message that is wordless, that is nothing more or less than the animals themselves—that the world is wild, that life is unpredictable in its goodness and its danger, that the world is larger than your imagination?"—Rebecca Solnit, *A Field Guide to Getting Lost*

Foreword

A peregrine falcon hunting in the sky above San Jose City Hall in search of prey. A porpoise swimming near Alcatraz. Sandhill cranes dancing in the Central Valley. A golden eagle soaring high above roadrunners and plovers in Panoche Valley. Sea lions and brown pelicans hanging out by the pier in Santa Cruz. A sea otter cracking open a clam on his chest in Monterey Bay. A small pika scurrying as hikers approach in Yosemite. An elusive mountain lion stranded in Griffith Park. Millions of monarch butterflies migrating up the Pacific Coast Highway.

This is my California.

Wildlife may not typically be the first thing that folks think of when they hear the word California, but throughout the nearly three years I lived out West I was amazed by the diversity of wildlife permeating every corner of the state. It wasn't what I was expecting.

When I was moving to the Bay Area in 2006, friends and family relayed wonderful stories about trips to San Francisco or Disneyland. They told tales about the horrors of Los Angeles traffic, their glimpses of movie stars in Malibu, or their confusion as they tried to find Silicon Valley on a map. Few of these tales included wildlife.

For many tourists and transplants, experiencing the Golden State means visiting its great cities and iconic structures. Thousands of people flock annually to places like the Golden Gate Bridge, the Chinese Theater, the Getty Museum, and Hearst Castle for inspiration and history, but many often miss the grandiose natural beauty of the state beyond maybe seeing the sun kiss Yosemite's Half Dome or taking a hike through the Seussian Joshua trees.

In contrast, my years in California (when I was working for the City of San Jose, helping former mayor Chuck Reed design and implement the city's Green Vision plan) were filled with breathtaking wildlife encounters across the state. I had amazing experiences at well-known wildlife hotspots such as Sequoia National Park, Elkhorn Slough, and Big Sur, but what also impressed me was the density of wildlife that I found in urban areas. I saw my first burrowing owls and gray foxes in San Jose. I remain on a quest to spot as many of the five hundred bird species that call Los Angeles County home for some part of the year. I was rendered speechless witnessing massive elephant seals battling at a roadside rookery at Piedras Blancas and spotting magnificent gray whales breaching from the shores of Half Moon Bay. It was as if the pictures from the dozens of *Ranger Rick* issues I had pored over as a child were coming to life before my eyes.

All across California, everywhere I looked, I found incredible wildlife.

On the pages that follow, my dear friend and colleague Beth Pratt-Bergstrom brings California's awe-inspiring wildlife to life. You will meet astounding animals and the dedicated people who are confronting immediate threats in hopes of securing their future. You will see how partners from a diversity of disciplines—including conservation groups, government agencies, private companies, and individuals—are coming together across the state for the common cause of restoring and growing California's wildlife populations.

Be forewarned, though, that this is not simply another book that you'll be able to read and then put down and forget about. This book is a treasure map of remarkable wildlife, much of which you can experience firsthand, and we believe that these stories cannot help but inspire you to act.

The truth is that we can each play an important role in ensuring a bright future for wildlife, no matter where we live, and these California stories serve as examples of how we can each take action, from anywhere across our country and around the world. The National Wildlife Federation works in every state and US territory toward organizing a truly inclusive, cohesive, broad-based conservation army consisting of state affiliates, public agencies, private land owners, and grassroots supporters of all stripes—all committed to restoring wildlife across America. Supporting policies and investments that advance conserving and restoring our natural resources are essential, but, as you learn about the remarkable species and work occurring all around you, it's my hope that you will also consider things that you can do as an individual to help wildlife.

We also hope that the stories in this book inspire you to help us in our goal to introduce millions of children to the wonders of wildlife. At a time when the average American child spends more than fifty hours a week glued to electronic devices (more than a full-time job!), it has never been more important for us to connect kids to the great outdoors. It's good for their health, their educational achievement, their critical thinking and leadership skills, and, most importantly, their general happiness.

I want all Californians and visitors to have the same experience with the state's spectacular wildlife as I've had the pleasure to enjoy. Reading this book is a great first step. Using this book as a treasure map to get outside and explore is even better. See you outdoors!

Collin O'Mara,
President and Chief Executive Officer
of the National Wildlife Federation

The Wild Wonder of California

"It was Deer on the Golden Gate Bridge! Hope they are safe! Kind of awesome!"—@Rydalia on Twitter

The Golden Gate Bridge stretches across San Francisco Bay, a vibrant orange-red band in a rainbow that includes the browns and greens of the Marin hills and the blues of the ocean and sky; it seems to extend from the natural landscape rather than be imposed on it. Underneath its steel arches, the ocean laps over the memory of land that reached almost thirty miles farther west during the last ice age. The bridge's massive towers serve as a fitting monument to honor what writer Rebecca Solnit described as "a drowned river mouth" where Columbian mammoths, ancient bison, and short-faced bears wandered fourteen thousand years ago. It seems to be the perfect ambassador for California, both an actual and theoretical bridge between a natural wonder and a manmade one.

Journalist Herb Caen once called the bridge a "mystical structure" possessing a heart and soul that "exerts an uncanny effect," and it even influences the weather, as its bulk can steer the direction of the city's trademark fog as it moves across the bay. It is the most photographed bridge in the world, and residents and tourists alike feel its irresistible pull as millions of people drive, walk, and cycle across the Golden Gate every year, or marvel over its visual echo from around the city.

Apparently, its draw is not lost on the animal world.

While I was working on this book, a story appeared in my news feed that caused me to exclaim aloud, "How perfectly wonderful!" The text read: "Two Deer Trotted across the Golden Gate Bridge Friday and It Was Magical." The story and accompanying video quickly went viral and became the top post that day on BuzzFeed, the *Huffington Post,* and other online news outlets.

As cars full of work-weary commuters packed the roads during the evening rush hour, two intrepid deer decided to do some exploring and make the almost two-mile trek across the bridge, stopping traffic as they journeyed from San Francisco to Marin. Like many tourists, they exited at the Vista Point parking lot, perhaps to take in the view of Alcatraz or Angel Island. As BuzzFeed remarked, "The deer were headed to Marin, which is known for its rugged forests and affluent liberal residents—so, not at all a bad place for a couple of does to spend a weekend."

Even if the deer didn't share the sentiment that they were embarking on a grand adventure, they inspired one for others. Even more magical for me was the scene captured in the many videos and photos posted by those lucky enough to be on the bridge at the time: three lanes of traffic patiently waiting behind the deer as escorts. Commuters who usually become annoyed at the slightest delay instead slowed down to snap photos and watch these wild creatures in their midst.

Rush-hour-traffic drivers celebrating a deer parade and wanting to ensure the animals made it safely across! That is the type of

Deer crossing the Golden Gate Bridge at rush hour.

story I aim to tell in this book: how people and wildlife are working it out in California in the context of a "new," reenvisioned
nature that challenges traditional notions of boundaries between
peopled and wild landscapes.

This isn't a science book—although you'll learn about the
natural (and cultural) history of the wildlife featured. As S. G.
Goodrich wrote in the preface to his *Illustrated Natural History of
the Animal Kingdom* in 1859, "I endeavor to reconcile something of
the sternness of science with the license of the describer, the narrator, and the anecdotist; I place myself between the Scylla of scientific naturalists on one side, and the Charybdis of popular taste
on the other." And this also isn't an exhaustive volume of wildlife
species or conservation efforts in the state. Please excuse my sins

of omission; captivating tales of wildlife, and what Californians have done and are doing for wildlife, could fill an encyclopedia series. Nor does this book address the bad news or what seem like the insurmountable problems animals face to survive, although these issues are important and what I work to combat daily. I am not asking that we ignore these issues, and indeed I cannot in my career. What I ask is that in reading this book you pause for a moment to remember what inspires us as well.

This book is simply a compendium of the wildlife stories that have captured my imagination, have elicited awe in myself and others, and that serve as examples of this new paradigm for how people and wildlife *can* coexist.

Most of all it's about wonder.

Although I have a science degree, my love of wild things originated not from learning scientific facts or figures but from being riveted to the movie *Born Free,* from whale watching on Cape Cod with my dad and feeling the ocean spray on my face from those magnificent creatures breaching into the sky, and from the simple discovery of a frog leaping from a tree limb in my backyard. The stories I've included in this book transported me back to childhood, to a time when I flipped through the pages of the supermarket editions of *Funk & Wagnalls Wildlife Encyclopedia,* eagerly awaited each new issue of *Ranger Rick,* watched Mutual of Omaha's *Wild Kingdom,* and simply reveled in the wonderment and magic of wildlife. In 2014, reporter Matt Weiser shared in *High Country News* his reaction to the appearance of the first wolf in California in ninety years. He remembered there was a "sense we were part of something miraculous," and he relished the gift of "wild wonder" that the wolf had returned to us.

Matt's term perfectly summarizes my book: these are stories of wild wonder.

Yet I don't write simply to spin a good yarn. My life's work involves conservation. And for wildlife to have a future in California—and the rest of the world—we need to foster a daily relationship with wildlife and nature. "There are some who can live without wild things, and some who cannot," said legendary conservationist Aldo Leopold. I argue that none of us really can. In California, where a mountain lion lives in the middle of Los Angeles under the Hollywood sign and deer trot across the Golden Gate Bridge, the reality is that no matter where you live—in a city, a rural town, or somewhere in between—we are all inextricably linked to the wildlife of our state, whether we regard them with awe, wonder, or fear.

I want to focus on the awe. My hope with this book is to start a viral meme of wonder for the almost forty million people that live in the Golden State—and beyond. Why? Because as Jake Abrahamson observed in an article for *Sierra Magazine,* "Awe prompts people to redirect concern away from the self and toward everything else. And about three-quarters of the time, it's elicited by nature." Quoting researchers, he wrote, "Fleeting and rare, experiences of awe can change the course of a life in profound and permanent ways."

I spend much of my time in awe encountering the wildlife that inhabits this state. I have the best job in the world as the California Director of the National Wildlife Federation. While overseeing the organization's conservation work in the Golden State, I continually witness instances of "wild wonder."

In Los Angeles, residents inspired by the plight of one lonely mountain lion are rallying to build what could be one of the largest wildlife crossings in the world. In Silicon Valley, the CEO of one of the most powerful technology companies in the world, Facebook, snapped photos from his office window of a family of gray foxes, posting to his thirty-one million followers that he's proud his employees made a home for wildlife in the middle of a high-tech campus. Porpoises cavort again in San Francisco Bay because of a grassroots effort to clean up a waterway that had once been a toxic mess. And after millions of residents gave a virtual cheer when learning that the first wolf in ninety years had crossed the state line into California, a crowd—including a toddler—later testified at a government hearing advocating for his protection. In Yosemite, people have taken responsibility for fixing a long legacy of the national park's bears suffering due to human transgressions. At the Hopper Mountain Wildlife Refuge, I held a magnificent condor in my lap, marveling over his rainbow-colored head and feathered muscled might—a relic from an ancient world, here because there are scientists dedicated to keeping this endangered species alive. Children are creating wetlands in their schoolyards to bring back California's fading native son, the red-legged frog; they're planting milkweed gardens to help save the imperiled monarch butterfly; and more than thirteen thousand people across the state have registered their backyards, schools, businesses, and places of worship as Certified Wildlife Habitats—transforming these places into welcoming spaces for animals.

As you'll learn in this book, animals have demonstrated they can sometimes adapt to human spaces. Can we adapt to wildlife as part of our everyday lives? The dominant philosophy of conservation has traditionally been to segregate people and wildlife—set aside islands of habitat—and although we must continue to do so, it's not enough. And it's not entirely working. Even in the best-protected places on the planet—national parks—some species are having a tough time. Nature has to be connected to work, and our cooperation is essential to creating and maintaining those connections that will ensure wildlife have a future.

When the number-one threat to wildlife worldwide is loss of habitat, we can no longer think of our cities or towns or neighborhoods, or even our backyards, as exempt from the natural world—or as off-limits to wildlife. We need to expand our view and realize that our shared spaces are as essential to conservation as our traditional protected lands. We need to create a new model of suburban and urban wildlife refugia.

This shift in thinking celebrates what I've come to define as the new paradigm for wildlife conservation: coexistence. And with this shift comes responsibility; wildlife today needs every one of us to lend a hand, not just researchers, environmentalists, and scientists.

People across the state are inspired by its wild wonder—inspired enough to take action—and this is why I am hopeful that California can serve as a new model of wildlife conservation. In the words of Rachel Carson, whose 1962 book, *Silent Spring,* ignited a new environmental consciousness: "The more clearly we can focus our attention on the wonders and realities of the universe, the less taste we shall have for destruction." This book is to remind us all about the remarkable wildlife that live in this state—sometimes in the most unexpected of places—and of the remarkable—and also sometimes unexpected—people who want them to thrive.

The Golden State is the birthplace of the modern environmental movement, but I don't want us to rest on our laurels. Wildlife conservation has entered a new age, forcing those who work to protect and preserve wild things to reconsider our approach. This "new nature" is personal, urban, social, and diverse, and the unique place that is California has given birth to it. The Golden State is known for spurring cultural seismic shifts (probably because it possesses the largest economy in the country), and it is full of contrasts, with enough superlatives to fill its own almanac. It's extremely diverse, both biologically and demographically. It boasts the highest and lowest points in the contiguous United States, and it houses every geographic region possible—ocean, mountains, deserts, and valleys. Called "one of the planet's richest places for plant and animal diversity," it contains the greatest number of total species and total endemic species of any state. As for its humans, California is the most populous state in the nation, with 90 percent of all residents living in urbanized areas. As of 2014, the majority of the state's residents are Latino, and 73 percent of Californians under the age of eighteen are people of color. California houses the largest number of national parks in the lower forty-eight states and also contains some of the largest cities in the country—both of which are essential to conservation efforts.

And so is the state's DNA of technology. Social media platforms are crowdsourcing a change in attitudes—and sometimes behavior—around wildlife, allowing us to become familiar with our animal neighbors in wholly new ways. My friend and colleague Leigh Wyman, who studies urban wildlife, sees our brave new world of social media as a major driver of the shift: "Social media is changing the conversation about wildlife; these stories have blossomed, propelled, and been shared, and a broad range of people are able to connect and have access in ways not possible before." Other technology, such as remote camera traps and GPS tracking devices, have also contributed to normalizing wildlife by revealing that our wild neighbors are all around us, and this knowledge helps rewrite our cultural norms and make it easier to have an ongoing relationship with the wild world. This urban, diverse, and crowdsourced movement represents a radical shift to thinking about wildlife as an integral part of our everyday lives, no matter where we live, and not just banished to the hinterlands or remote wild areas, maybe seen once a year during a vacation to Yosemite. It's not about habituating wildlife, it's about habituating ourselves to the wild world.

We can have daily relationships with a mountain lion like P-22 or the wolf named OR-7 because they can now—with a little bit of help—post selfies to their Facebook pages like the rest of us. This is California's cult of celebrity applied to the animal world, and put to good use, for these are reality shows worth watching because they can have significant impacts on conservation. Jon Mooallem, the author of *Wild Ones: A Sometimes Dismaying, Weirdly Reassuring Story about Looking at People Looking at Animals in America*, made this point full force in a recent TED talk:

In a world of conservation reliance, those stories have very real consequences, because now, how we feel about an animal affects its survival more than anything that you read about in ecology textbooks. Storytelling matters now. Emotion matters. Our imagination has become an ecological force.

Through these tales of wild wonder, I ask what is possible, focusing not on what we've done or continue to do wrong but on what we have done and can do right. As Marie De Santis states in her delightful book *California Currents: An Exploration of the Ocean's Pleasures, Mysteries, and Dilemmas,* "If after using this book you have more questions than answers, then its purpose is served—to heighten your sense of wonder." I don't have all the answers, but I can pose possibilities—some attainable, some admittedly wishful thinking but with good intent. We still have a lot of work to do to ensure wildlife has a future beyond all the challenges they face. But I hope these stories will inspire you to think more about our wild neighbors and what you can do to help, whether it be building a landmark crossing over a ten-lane freeway, planting milkweed in your backyard, or simply waiting patiently in traffic to give wildlife a chance.

Steve Winter's famous photo of P-22 in front of the Hollywood sign.

CHAPTER 1

A Mountain Lion in Hollywoodland

CAN PEOPLE AND WILDLIFE COEXIST
IN THE SECOND-LARGEST CITY IN THE NATION?

"The mountain lions have not learned, like the wolf, to get the hell off the land....They have an odd, powerful dignity that does not understand the endless catches and snags of the human race. That is why they are still at the fringes, still fighting."—Craig Childs, *The Animal Dialogues: Uncommon Encounters in the Wild*

"All this talk about #sharkweek. Whatevs. Call me when a shark moves into a huge park in the middle of a city."—@P22MountainLion, a P-22 Twitter account

Imagine the second-largest city in the country asleep. Or if not asleep, as close as the City of Angels will ever come to slumber. The dark obscures the city's own north star—the forty-five-foot-tall letters of the Hollywood sign—and a parade of club-goers cruise home along Sunset Boulevard, some making one last stop at In-N-Out Burger before being swept up in the unrelenting pulse of highway traffic. This is, after all, the city that termed a weekend freeway closure "Carmageddon."

For the metropolitan area of Los Angeles, home to almost thirteen million people, the deep hours of the morning provide as quiet a time as it ever experiences. The Spanish named the hours between midnight and dawn *madrugada,* a time when most of us are lost in dreamland. But one would-be resident of Los Angeles is awake and alert. He strolls through the magical time of the madrugada thinking not of fame or riches but something far humbler: deer.

This newcomer is a mountain lion.

A relative youngster at two years old, P-22—as he will soon be known to the world—heads east toward the city, having probably come the twenty miles from Topanga State Park in the Santa Monica Mountains. He walks regally, muscles rippling beneath his tawny coat, that unmistakable long tail twitching at times. His leaving home, called "dispersal" by biologists, marks a typical milestone for a cougar at his age, as young males must seek out their own territory. The Santa Monica Mountains house plenty of deer, but those deer can come with a high price if located in the established home range of another male. Cougars can fight to the death over territory, and a teenager like P-22 knows he is no match for an older, more experienced cat.

Finding unclaimed space that includes a deer herd within a wilderness squeezed on all sides by a megalopolis can prove to be challenging. P-22, however, fully utilizing the stealth that he has inherited from his ancestors over millions of years, saunters unnoticed through the neighborhoods of Bel Air and Beverly Hills, his paws leaving impressions on the impeccably manicured lawns.

Mountain lions evolved in the Americas and, unlike their cheetah cousins during the last ice age, did not venture across the Bering Land Bridge to Eurasia in search of food. Why? Some scientists have theorized that they were reluctant to follow their kin because they are "ghost cats," averse to open areas and plains. So perhaps we owe the unique presence of this top predator to a case of collective agoraphobia? That evolution has shaped one of North America's largest carnivores into a shy, introverted, and enigmatic creature is not without its irony, as Craig Childs notes in his book *The Animal Dialogues*: "So now the biggest, most dangerous animal is also the quietest and the hardest to see."

Yet given the elusive nature of cougars, why does P-22 chart a course into the most crowded area in the United States?

Scientists will provide a basic answer: he is searching for an unoccupied space with food. While this explanation is not technically wrong, it seems too simplistic to reduce P-22's journey to a routine trip to the grocery store; something else must be at play in this adventure, something beyond the need to escape the wrath of other lions and to answer the demands of the stomach. There are easier ways to secure deer than to march into the middle of Los Angeles.

As P-22 wanders through the neighborhoods featured on Maps of the Stars—possibly stopping to nap among the oak trees at the Mountain Gate Country Club, its members teeing off unaware that a 120-pound cat dozes within swinging distance—what guides him? What urge keeps him heading east despite the constant human presence he encounters? How does he know his own personal deer park—where he can reign like a medieval ruler—exists miles away, past the maze of congested streets, past the hillsides with every inch of green blotted out by seemingly endless sprawl?

Maybe P-22 is following the abundance of deer trails—invisible to us—that crisscross the subdivisions and golf courses. Perhaps the revived flow of the Los Angeles River, slowly being daylighted after having largely been buried in concrete for fifty years, is once again singing the promise of prey and enticing a top predator into the city limits.

Scientists warn us about anthropomorphizing, but if we share 92 percent of our DNA with a mouse, it seems ridiculous to keep asserting that animals operate solely on instinct and nothing more. Dolphins recognize themselves in mirrors, rats display empathy, and elephants hold funerals to mourn their dead. Would it really be a stretch to say P-22 possesses a sense of adventure?

Researcher Kerry Murphy, who has logged countless hours observing cougars, told Chris Bolgiano, the author of *Mountain Lion: An Unnatural History of Pumas and People*: "What's impressed me most is their individuality. Each of them has a life that's as real to them as ours is to us."

P-22 just may be the Neil Armstrong of his kind. A quick glance at his route on a map shows he had to be a bit mad to even attempt his journey. To get to his new territory of Griffith Park, he must cross two of the busiest freeways in the United States.

Imagine soft, padded paws fitted for bounding over snow and boulders touching the asphalt of the first eight-lane highway, known as one of the worst roads in the country. Even in the

P-22's journey to Griffith Park took him across two major freeways.

middle of the night, the 405 never slows, and the highway thrums with mechanical noise and explodes with the mad dance of headlights. When faced with the living, breathing monster of the 405, most cats do an abrupt about-face, or get mangled by a few tons of moving steel. But P-22, with his tenacity, or luck, or both, somehow manages to cross. There is no way of knowing how he navigates the formidable obstacle of the road, whether he uses an under- or overpass or bolts straight across. All have been attempted by other cats, and many haven't lived to tell the tale.

My guess? He probably did what most of us do when confronted with the Los Angeles freeways: floor it and hope for the best.

Imagine that bound! One large step for cougarkind. Mountain lions can jump a span of forty-five feet. Someone might have seen P-22, startled by the view of him dashing across the road in a blur of maniac motion, but since mountain lions are not a usual reality for the LA motorist, that long tail or autumn-brown coat in the headlights was probably attributed to a large dog.

His miraculous feat, however, only pushes him into more densely populated areas, where he must keep going, perhaps thinking to himself he has hit the point of no return. I imagine his final miles as akin to a thirsty man wandering in a desert, hoping for signs of water with every step. Then, like a mirage, the Hollywood Hills appear, a green expanse filled with deer. Even more importantly, he senses no indication of another male lion. Cougars leave scent marks of urine or feces, or scrapings on trees, to designate their territory and warn other lions to keep out.

One last push. He might stand on the Mulholland overlook at night, gazing at the city lights of downtown to the south and the lack of lights on the landscape due east, another promising sight. He might consider his options for crossing the 101, peeking out of his hiding place while he rests during the day, smelling the heavy stench of gasoline and exhaust, the noxious perfume of the freeways. Perhaps he's also curious about those giant white letters jutting out of the hillside.

P–22 somehow navigates the 101, ranked by some as the worst commute in America. He might pause a moment with a triumphant look back at the speeding cars, then pick up his pace for the last half mile to his destination, sauntering through the winding roads and quiet neighborhoods, taking note of the Hollywood Reservoir, a place he will soon frequent. If cougars feel relief, I am sure P–22 does at this moment. No houses. No other male cougars. Plenty of deer. #winning

And then he creates a marking as significant in the cougar world as the famous boot print on the moon: he scrapes a tree with his claws, forcefully and with much satisfaction, and claims Griffith Park for his own.

"On February 12, 2012, at 9:15 p.m., we collected the ultimate evidence....The discovery of this mountain lion remains one of my proudest moments as a wildlife biologist and as an Angelino who grew up on the edge of Griffith Park."—biologist Miguel Ordeñana

Griffith Park represents a small fraction of a normal home range for a *Puma concolor*. An adult male can occupy up to 250 square miles; P-22 somehow makes do with 8—the smallest home range ever recorded for an adult male mountain lion. His incredible journey of dodging semitrucks and sneaking through country clubs inspires awe, but his ability to survive like a castaway stranded on an island surrounded by an ocean of city ranks in the category of incredible. And for P-22 there is not much hope of a rescue boat. He'll have to swim through more traffic to escape.

At more than four thousand acres, Griffith Park ranks as the nation's largest municipal park that houses wilderness. It is a hybrid of city and nature surrounded by a spider web of freeways, only two miles from the Hollywood Walk of Fame. Ten million people visit the park annually, more than double the visitation to Yosemite and Yellowstone. Within its borders they can play a round of golf or a set of tennis, visit the famous Griffith Park Observatory, attend concerts at the Greek Theatre, ride the historic merry-go-round or miniature train, cheer on youth soccer or baseball, take their kids on pony rides, canter their horses along the equestrian trail, and picnic, hike, bike, or run.

A Nicaraguan American who grew up in LA, Miguel Ordeñana spent his childhood playing hide-and-seek among the oak trees and hiking with his family up to the Hollywood sign (when you still could). Grown-up Miguel is one of those quiet and unassuming individuals whose calm manner should not be mistaken for a lack of passion or ambition. He has studied bats in the Mojave Desert and jaguars in Nicaragua, proof that a childhood in the heart of a large city can still foster a love of wild things—or, argued another way, proof that nature does exist in Los Angeles despite our best attempts to banish it. "Griffith Park was my first wilderness," he says. "I didn't know the difference between that and a Yosemite. Now I know the difference doesn't matter, at least to a kid discovering nature. Griffith Park is more accessible and just as meaningful."

Miguel's employer, the Natural History Museum of Los Angeles County, has a mission to dispel the notion that LA is an urban wasteland, instead promoting the city as a wildlife hotspot. This quest isn't as improbable as it sounds. With more than five hundred avian species recorded in the county, Los Angeles ranks as "the birdiest county in the United States," and its diverse geography of ocean shoreline to mountain peaks houses an immense range of biodiversity. But the museum's initiative is not just about gathering anecdotal superlatives—it is conducting the world's first and only long-term study of urban biodiversity.

The museum's new interactive and technologically sophisticated Nature Lab and Nature Garden celebrate the city's urban wildlife through video games, real-time mapping, an "opossum cam," and an outside habitat complete with raptors, dragonflies, and raccoons—wildlife liberated from the stuffed-animal dioramas of the past and reimagined for the younger generation. This is wildlife for the social media age: coyotes in video games, live critter cams, and people taking selfies with rattlesnakes.

As part of the Griffith Park Connectivity Study, Miguel is a member of a team that includes Erin Boydston (US Geological Survey) and Dan Cooper (Cooper Ecological Monitoring, Inc.) and that collaborates with Laurel Serieys (a PhD student at UCLA and founder of the website Urban Carnivores). They have contributed to a long-term study, led by National Park

Service biologists Seth Riley and Jeff Sikich, on the impacts of urbanization on cougars, bobcats, and other wildlife.

Miguel has monitored wildlife movement between the park and the Santa Monica Mountains to the west since 2011. For most of Griffith Park's history, the city managed it almost entirely for the needs of human recreation, and not until 2007, long after some native plants and animals had already gone extinct from the area, did anyone conduct a formal survey of its biodiversity. Local residents, such as Friends of Griffith Park founders Gerry Hans and Mary Button, became tired of the inattention paid to the park's flora and fauna and decided to do something about it. "We wanted to give the wildlife a voice before it all disappeared," Gerry said. They helped fund the connectivity study, key to the future of the coyotes, deer, bobcats, foxes, and other wildlife marooned on the island of green. By using remote cameras, the team investigates how wildlife navigates (or doesn't navigate) the barriers that surround the park, and this includes monitoring crossing sites along the 101, 5, and 134 freeways.

Remote cameras are an important dimension to wildlife field research, as they are allowing scientists to witness behavior that either would have never been observed or would have required thousands of hours of field study to understand. In fact, given P-22's ancestors' gift of stealth, without this new technology the big cat might have roamed in Griffith Park for years without anyone knowing it.

The job title Wildlife Biologist sounds romantic and exotic—visions of Jane Goodall hugging chimpanzees come to mind—but in reality it now involves a significant amount of time staring at computer screens. For the Griffith Park crew, sorting through the thousands of photos acquired each month from the remote cameras can be a tedious job. In February of 2012, while flipping through yet another batch, Miguel, fatigued with the flow of coyote and deer images, clicked the mouse in autopilot mode.

Coyote, coyote, deer, squirrel, cougar…

A portrait of P-22 in November 2014.

Cougar? That jolted him out of his boredom. Were those the hindquarters of a mountain lion?!

As Miguel remembers, "I gasped and stared for a while, astonished at the size of the animal's tail, body, and paws. I went through the photos again...trying to see if my mind was playing tricks on me—maybe it was just a Great Dane that got loose late at night that had stood very close to the camera. But I immediately knew what it was. Then I thought, 'You should probably call somebody.'"

Miguel called his partners on the Griffith Park study and the National Park Service biologists and left a series of excited and frantic voice messages on their cell phones. Receiving the news of a cougar in Griffith Park was akin to hearing they had won the lottery. Though all of these scientists could probably be studying wildlife in more exotic locales like Yellowstone or South America, they have become enchanted with urban wildlife and dedicated their work to ensuring that it thrives in LA.

The researchers are the sort of people I like to hang around. They keep dinner conversations lively with their tales of gruesome scavenger hunts for carcasses, performed to monitor the diets of carnivores. You can't help but have an affection for people who get excited about things like disemboweled raccoon remains and the dissection of bobcat poop. When Miguel and Laurel stumbled upon evidence of P-22 eating a coyote, their exuberance clearly showed in the blog post Miguel published about the discovery—"What's on P-22's Menu"—complete with the gory photos. One line reads: "Laurel shouted out to me with excitement!!! 'It's a coyote!!'"

What is *on P-22's menu?* many ask, either out of fascination or fear. For the answer, we can thank these devoted researchers, who bushwhack through dense chaparral and poison oak in order to kneel beside pungent rotting meat for insight into the cougar's palate. What the National Park Service study has told us to date is that you can take the mountain lion out of the country, but you can't take the country out of the mountain lion. P-22 eats almost exclusively deer, and the team has found no evidence of his making a meal out of a house pet. As Miguel observes, these results "dispel myths about urban mountain lions seeking out pets or becoming dangerously habituated to human-subsidized food resources. P-22 is retaining the same natural behavior of his more rural counterparts and going after deer and other natural prey in the wildest patches of the park."

If deer can be said to possess a nemesis, it's the mountain lion. All of its muscle and might are dedicated to the pounce and takedown of deer, and a cougar's "lithe and splendid beasthood," as described by wildlife author Ernest Thompson Seton, seems wasted on smaller prey. Considering the energy required to make a kill, a raccoon or a house cat is hardly worth the lion's effort.

When deer are scarce, however, cougars are opportunistic killers, especially since they can't nibble on berries or acorns. Unlike coyotes or bears, a mountain lion is a true carnivore and rarely, if ever, consumes vegetation, as its digestive system rejects it. In the animal's ongoing quest for meat, evolution has provided the right tools for this highly specialized ambush predator. Powerful haunches propel a cougar into leaps of up to fifteen feet and can launch short sprints up to forty miles per hour. Even the teeth—canines that can grow as long as an inch and a half—possess a secret weapon: an abundance of nerves to help the cat sense when it has penetrated the precise kill point.

Faced with this formidable predator, the deer have two defense mechanisms: running and hiding. Deer in Griffith Park must have made for an easy meal at first, as many of the prey animals in Griffith Park had entirely forgotten the long-absent predator. P-22 quickly reawakened their memory. He kills about three or four deer a month, and the evidence of his carnage signals to the researchers a promising development. As Miguel points out, "These grisly scenes also may provide a sign of hope for Griffith Park remaining a functional urban oasis"—meaning the appearance of apex predators like mountain lions are a good indicator of the overall health of an ecosystem, sort of nature's thumbs-up that all systems are a go.

To call the park an oasis, however, implies it is a way station, a place where a weary traveler can stop and refuel before moving on. P-22, unless he attempts once again the perilous journey across LA's freeways, has nowhere to go. Although the all-you-can-eat deer buffet might be able to sustain him indefinitely, his solitary existence may not.

P-22 needs a mate. And for that he must travel.

"In addition to her youth, good looks, independent spirit, love of the great outdoors, and proven ability to bring home the bacon, P-23 does not appear to be related to P-22. So what's keeping these would-be soul mates apart? The 405. Yes, the east/west LA divide has defeated many a romance before, but knowing what we do about these two, we think they can defy the odds. Go get her, lion! (But safely, please.)"—Shayna Rose Arnold, *Los Angeles Magazine*

P-22 has won over the heart of Los Angeles. Reporter Martha Groves of the *LA Times* first broke the story, and the lion has been a media darling ever since. He has his own Facebook page ("P-22 Mountain Lion of Hollywood") and several Twitter accounts, where he tweets requests for good restaurants serving "raw meat and hikers," bemoans the perils of the 405, and claims he ate former city council member Tom LeBonge. His bachelor status rallied Angelinos to his plight, and he might be the first cougar in history with a dating site and an entire city playing matchmaker. When a photo of P-23, a young female cougar, taking down a deer on Mulholland Highway went viral on social media, many a blogger, like *Los Angeles Magazine*'s Shayna Rose Arnold, thought she had found P-22's ideal mate.

Biologist Jeff Sikich is also concerned with P-22's love life, though his interest as a National Park researcher relates more to genetics, inbreeding, and connectivity issues than romance. He and Seth Riley, the foremost expert on urban carnivores, have collaborated on a long-term cougar study in the Santa Monica Mountains National Recreation Area since 2002.

Tracking and studying lions in Los Angeles—and beyond— is Jeff's occupation. He has built an impressive resumé that includes studying big cats in the Peruvian Amazon, Indonesia, and South Africa, and scientists all over the world fly him to their locales so they can learn his "safe capture" technique for big cat research—modified padded foothold cable restraints that send immediate text message alerts to the scientists monitoring the traps. As the *LA Times* said in a profile, "Sikich's instincts in the wild and his humane captures have earned him a place among a cadre of go-to carnivore trackers." Yet despite his global renown, his heart remains with the Los Angeles lions.

When I first meet Jeff, he stands in a grove of live oak on the edge of a parking lot adjacent to the famed Griffith Observatory. In the movie of P-22, Jeff would play himself, as his rugged good looks and affable, humble manner perfectly fit the casting call of "hero biologist." People casually walk or jog by, consider the decidedly science fiction–like radio antenna he's holding in his hand, and either pass it off as another Los Angeles eccentricity or a scene being filmed for a movie, a sight to which Angelinos are largely indifferent unless someone with real star power is involved.

Some passersby, however, are curious enough to ask what in the heck he is doing.

"Tracking a mountain lion," he replies.

Probably not the answer most expected. Of those who inquire, not a single person—even the woman walking two adorable but vulnerable corgi dogs—expresses fear. Responses range from "Cool" to "Wonderful that he is here" to "Can I see him?"

Public relations is part of the job when you are studying lions in a city of millions, and Jeff is a gracious and witty spokesperson for cougars. People truly want to know more about this cat, from his height and weight to their likelihood of being mauled. And without Jeff's talents, we would not know enough about P-22 to answer these questions. He's the one who captured the cat and fitted him with the radio collar that tracks his movements. He is also the one who named him. Although some advocate for what they consider a better moniker (a recent contest elicited suggestions including Pounce de Leon, Griffy, and Puma Thurman), P-22 appears to be growing on people. As part of a numbering system for tagging animals, P represents "puma," and 22 denotes his sequence in the number of cats that have been tracked in the Santa Monica Mountains.

Jeff has spent intimate time with P-22 on multiple occasions. When a failed GPS device necessitated his recapture, Jeff, quite remarkably, simply snuck up on the cat. After picking up residual transmitting signals, he spotted P-22 crouched in a ravine, at which point he climbed an overhanging limb and darted the cat with tranquilizers from ten feet away. "He knew I was there and made no move to attack," he recalls. "He was probably hoping his strategy of concealment would keep working. This shows you the lengths these cats will go to avoid a human encounter."

In the world of cougar research, captures are usually made with hounds treeing an animal, then the researcher tranquilizing it. In an urban area, however, you can't have dogs chasing a cougar into someone's backyard gazebo or up their apple tree. Jeff improved upon an existing solution—the padded foothold method—which lures the animal into the trap usually by playing a recording of a female in heat or a deer in distress. Once the animal springs the trap, the system issues an immediate text message to researchers, minimizing the animal's time in the padded foothold. The technique isn't foolproof, and there is no question that using the hound method, along with cage traps—both of which he does whenever possible—would make Jeff's life easier. As he puts it, "This is an animal that can roam 250 miles and we're trying to get him to step into an area the size of a dinner plate."

Once, when Jeff needed to replace a failing collar on the cougar known as P-10, he tracked the cat to his hiding place in a bushy area right outside a garage in the residential area of Pacific Palisades. He knocked on the door, assured the homeowner that she "had no reason to be alarmed," and then informed her a mountain lion was napping in her yard. "Do whatever you need to do," she responded. "I love kitty-cats."

After Jeff blow-darted P-10 with tranquilizers and took the animal's measurements, the homeowner brought him a sandwich

The world's most famous mountain lion, P-22, coughing up a hairball.

and a cold drink. He then allowed her and a neighbor, Bill Fado, to get a photo of the tranquilized cat. Bill recounted the once-in-a-lifetime experience in the *Palisadian-Post*: "As I watched P-10 slowly walk away from us, occasionally looking back, it brought to mind images of the plains of Africa. Then he was gone, or so we thought. Suddenly, there he was, 50 feet away looking right at us. He stared at us for 10 seconds as if to say, 'Thanks for getting me out of that jam,' then turned away and went off into the night. It was a mystical experience that I will never forget."

Mountain lions in neighborhoods make for good stories, but they prove to be the exception to the rule. What Jeff and Seth's research demonstrates is that for all their urban living, cougars actually don't like to spend much time near people. Of the more than thirty thousand GPS readings from the first eight cats in their study, 98 percent occur in natural areas and two-thirds register greater than a kilometer away from urban spaces. So it's likely that even if P-22 kills a deer near a house or a high-use recreation area, he will hide or cover it as cougars typically do, then retreat for the day to a remote part of Griffith Park until he can return for solitary midnight snacking.

Snacking is his main business. As Jeff jokes, "He's spent his time like Rocky. Getting in shape, fattening up. If this was a normal situation, he would probably go somewhere to challenge a male or take over a dead male's territory." But so far, P-22 has been content to stay put, with only occasional forays into the neighborhoods surrounding Griffith Park, such as when he was discovered napping in a crawlspace in the Los Feliz home of Jason and Paula Archinaco. "We have three cats, and I don't see why we can't make room for one more," Paula told the press—because of course there was press for a Hollywood celebrity of his caliber. The resulting media circus, complete with a live feed of the cat peeking out at the cameras and helicopters buzzing the neighborhood, made headlines around the world. (*Time* magazine announced: "The Mountain Lion that Was Hiding Under a Los Angeles Home Has Left.") The media spectacle was not born of fear, however; for the most part, people just wanted a glimpse of the famous cat.

As P-22 has proven, we don't need to unthinkingly fear mountain lions. Even when trapped in the crawlspace, surrounded by a crowd of people, and having tennis balls air-gunned at him in an attempt to haze him out of the house, he made nary a threatening move. As to his harming a human, Jeff underscores that attacks by cougars are very rare; the chance is about one in twenty-five million. You're more likely to win the lottery or be struck by lightning than be killed by a cougar. "These are large carnivores capable of attacking people, and they deserve a healthy respect," Jeff says, "but clearly mountain lions don't think of people as prey, and this is good news for both people and lions. If they wanted to eat us, they would."

Jeff goes on to say that in Los Angeles, for both P-22 and people, the freeways pose the greater risk. In Los Angeles County alone, automobiles cause on average about 750 deaths and 85,000 injuries each year. In California, mountain lions have attacked fewer than twenty people and killed just three since 1986, according to the state's Department of Fish and Wildlife. These statistics don't diminish the tragedy when a person is killed or injured by a lion, but it puts the risk in perspective. Living in lion country is much safer than living in car country.

The automatic fear—and the killing of cats that simply appear on the urban landscape—is usually due to a lack of education, not maliciousness. If you're not a carnivore biologist and accustomed to mountain lion behavior, then a hissing, snarling, 130-pound cat can easily lead to a state of panic. Simply learning about normal lion behavior would help dispel some of the fright and help people realize that the majority of encounters with lions end without incident. As the California Department of Fish and Wildlife's Lt. Kevin Joe notes: "Just because you find a mountain lion behaving normally, but in an unusual location, it doesn't mean it's a threat to public safety."

How long can or will P-22 keep up this duck and cover, this solitary and stealthy existence in the middle of LA, before he gives up his home for the urge to mate? During one of our drives around Griffith Park, I asked Jeff about P-22's options. Is it possible he might remain a bachelor forever and just stay where he is? Jeff considers it. "This cat has already done things we've never seen, so anything is possible," he says.

The other scenario many have proposed—for the safety of both the lion and his human neighbors—involves relocating P-22. Jeff appreciates the sentiment and the logic behind the idea yet doesn't believe it's a good option. "Rarely is relocation successful for males," he says. "The lion might just come back or get killed doing so. You are dropping a cat into unfamiliar territory and probably forcing an encounter with an older or more experienced lion, which usually results in the death of one of the males."

But the more likely outcome—that he heads out on his own—probably won't produce a very happy ending either. "I don't know what will happen if he leaves," Jeff says. "He did find a safe way across once, but chances are slim that he will make it past the free-ways again."

People are genuinely concerned for P-22's safety. When staying in Thousand Oaks for a briefing about the National Park Service's mountain lion study in the Santa Monica Mountains, I arrived very late to the hotel, and an obviously bored but very friendly night clerk checked me in. She gave me directions to the nearest Starbucks, then asked why I was in town. I told her the story of P-22 living in the middle of Los Angeles. She looked at me, fear-stricken for a moment. I prepared to launch into my usual speech about how you are more likely to be struck by lightning than attacked by a mountain lion when she said, "He's safe there, right? No one is going to hurt him?"

This affection for mountain lions isn't cursory. If you live in California, the odds are almost fifty-fifty that you live in mountain lion habitat. In 1990, California residents passed Proposition 117, known as the California Wildlife Protection Act, which reclassi-fied the lion as a "specially protected mammal" and banned the hunting of lions in the Golden State—this even though mountain lions are neither endangered nor threatened in most of California. California is the only state to date that has banned the hunting of lions. Admittedly, not every resident of the Golden State expresses awe and wonderment at seeing the trademark flick of a cougar's tail, yet even the rare attack does not deter the unwavering sup-port of the majority, who enjoy having cougars on the landscape. P-22 prospering in Hollywoodland serves as just one example of Californians' affection toward the cat.

Unlike in most of the state, the cougar population in the Santa Monica Mountains is imperiled. The survival of P-22, and indeed the survival of all the cats in the Santa Monica Moun-tains, largely depends on one factor: connectivity. Translated into non-biology terms, they need to be able to get across freeways and roads from one natural area to another. The superinten-dent of the Santa Monica Mountains National Recreation Area,

David Szymanski, started a public briefing about P-22 reinforcing the need for vital linkages: "If the lion didn't exist as a poster child of the importance of connectivity of open space, we would have to invent him."

As much as we like to play matchmaker for P-22 and impose our thoughts of romance upon him, the biologists are not so much concerned about his loneliness as they are invested in his role in increasing the genetic diversity of the lion population. Inbreeding leads to birth defects, such as kinked tails, weak spines, and single testes, and the fewer viable males you have in a population, the faster it implodes. With animals unable to reliably cross highway barriers, and automobiles taking out most who even try, the cats of the Santa Monica Mountains display some of the lowest genetic diversities of a lion population in the West. The researchers look to the almost extinct Florida panthers as a low bar they do not want to reach. Ultimately, if these cats can't move and find mates from other areas, the entire population is at risk, not just the charismatic P-22.

"The Santa Monica Mountains alone just cannot support a viable mountain lion population. There just isn't enough room," says Seth Riley, Jeff's partner in research. Yet he is optimistic. "Personally I am hopeful despite the challenges. They continue to survive naturally. But landscape connectivity is key, and it's pretty darn bad right now across the 101."

Seth is one of the leading experts in the science of urban cougars and other wildlife, and he literally wrote the book on the subject (*Urban Carnivores,* with coeditors Seth Riley and Stanley Gehrt). He becomes animated when he talks about what he considers the ultimate goal of his and Jeff's research: a wildlife crossing along the pinch point they have identified on the 101 at Liberty Canyon, a key passage for mountain lions and other animals. "This is a vital crossing in one of the last undeveloped areas on the 101, and building a safe passage gives us a chance to ensure the future of the cougars—and all wildlife—in the Santa Monica Mountains and Los Angeles area," he says. Seth's research has also shown that an array of wildlife—even smaller creatures like salamanders—are impacted by these paths.

Jeff and Seth took me on a tour of the site and pointed out where the crossing would be located. We hiked up a hillside and from our vantage point could easily see the undeveloped land funneling toward the highway on both sides into an animal dead end. I imagined a bridge stretching across the 101, and the first tentative footsteps from a bobcat or mountain lion, or even a salamander. It's a massive undertaking that comes with a price tag in the tens of millions, and public support is essential. Christy Brigham, the National Park Service's chief of resources, underscores this last point: "We are not going to be able to keep lions in the Santa Monica Mountains unless we all think it's a good idea."

P-22 probably also stands on a hillside at night considering his options to leave Griffith Park. He's lived here since 2012, and is long past the age a mountain lion settles down. His fate is uncertain. He might decide to remain a bachelor and spend the remainder of his days in the city. He is still vulnerable to human threats, like poison exposure (see page 29) or being hit by a car. His allegedly having made a meal of one of the Los Angeles Zoo's koalas in March 2016 prompted some calls for his relocation, but supporters rallied against his eviction. Even the zoo itself has been forgiving; a headline in *The Washington Post* read "L.A. Zoo to the mountain lion that probably ate its koala: No hard feelings." He might view with longing the mountains of the Angeles National Forest, less than ten miles away as the crow flies, nothing much for a cat used to traveling twenty-five miles a day. But even the most direct route would entail crossing the 5 and two other freeways, and traversing the city streets of Burbank or Glendale.

When I ponder the plight of P-22, I conjure up an image of Los Angeles shutting down for a day. I picture the cat sensing the unprecedented quiet, sensing that the monotonous noise of the cars has ceased, and then sprinting up Highway 5 with onlookers cheering his progress. (Is a ticker tape parade too much to add to the fantasy?) They watch as he heads north, perhaps as far as Los Padres National Forest, where he could lose himself in the almost three thousand square miles of protected areas and leave his freeway-cruising days behind him.

Now, shutting down freeways is unrealistic, but building a bridge for cougars and other wildlife is not. The Liberty Canyon Wildlife Crossing is becoming more of a reality with the launch of the National Wildlife Federation and its partners' #SaveLACougars campaign, which has rallied thousands to work toward the bridge's completion. P-22 has largely inspired this effort. Building the largest wildlife crossing in the world in the most densely populated urban area in the United States would send a message to the world. For a city that has long been the poster child for environmental degradation, P-22 could be the tipping point, and a chance for redemption.

One proposed alternative for the Liberty Canyon Wildlife Crossing, by senior architect Clark Stevens, of the Resource Conservation District of the Santa Monica Mountains.

"Then it suddenly occurred to me that, in all the world, there neither was nor would there ever be another place like this City of the Angels. Here the American people were erupting, like lava from a volcano; here, indeed, was the place for me—a ringside seat at the circus."
—Carey McWilliams, *Southern California Country* (1946)

In 2013 Kathryn Bowers, coauthor of the provocative and fascinating book *Zoobiquity: The Astonishing Connection between Human and Animal Health,* invited me to speak on a panel for Zócalo Public Square called "Does LA Appreciate Its Wild Animals?" Just a few years ago, my answer to that question would have been a resounding no.

Los Angeles didn't come natural to me, as I suppose it doesn't come natural to most of us. I have since changed my mind, and my conversion nearly echoed that of Carey McWilliams in the 1940s. He shares in his book *Southern California Country: An Island on the Land* that the city at first appalled him. But then, after developing a begrudging respect over time, he suddenly realized he wanted to be a part of this explosion, the new, emerging "lovely makeshift city."

For me, a young girl who yearned to live and work in national parks and then eventually achieved that dream, I once considered Los Angeles an abomination, a place that, as author William Deverell described, "willed itself by shoving nature around." It was P–22 that lured me and P–22 that allowed me to see LA in a new way. If LA's version of nature is good enough for P–22, then who am I to judge? If the city can support a 120-pound predator, it can also provide homes for foxes, bobcats, birds, insects, reptiles, and amphibians.

Miguel Ordeñana agrees: "It didn't take me too long to recognize the scientific and conservation significance of P–22's story, and the media coverage helped me learn that he was going to be a special ambassador for Griffith Park, LA wildlife, and urban mountain lion conservation. Knowing that there is a mountain lion in LA's most popular and accessible park provides a bold statement that there is plenty of nature to explore even in urban Los Angeles."

He's right. LA's 468 square miles of land and 34 square miles of water extend to the Santa Monica Mountains and the Pacific Ocean, include nine lakes, one river, and a million trees. Within its borders are 390 public parks and 15,710 acres of parkland.

Surprised? Most people—even some who live in LA—are not aware of the immense connection the city still retains to the natural world. Los Angeles has made nature its own, woven its own unique cultural landscape onto the physical one, and perhaps shaped the tale of Mother Nature into a structure it's comfortable with—that of a Hollywood blockbuster screen-play. Jenny Price, author of the brilliant essay "Thirteen Ways of Seeing Nature in LA," writes: "The history of Los Angeles storytelling, if more complicated, still basically boils down to a trilogy. Nature blesses Los Angeles. Nature flees Los Angeles. And nature returns armed."

Isn't P-22's improbable story the stuff of which blockbusters are made? It's the puma version of *Star Wars,* or *Rebel Without a Cause*—a restless young man with a troubled past comes to a new town, finding both friends and enemies—with maybe a bit of *The Big Lebowski* thrown in. And what better setting than Griffith Park, where scenes from *Rebel* were actually filmed and where today a statue honoring James Dean stands next to the observatory.

Los Angeles has been deemed a "land of magical improvisation," and in this new zeitgeist of urban wildlife relationships, it seems to be fulfilling this description as well. As LA city wildlife officer Greg Randall offers, "Los Angeles is wildlife habitat with houses on it." Shifting our perspective to that view opens up possibilities beyond thinking the coyote a villain and the mountain lion a monster.

In a state where 90 percent of people live in urbanized areas, where wildlife is running out of space, and where people are becoming increasingly disconnected from nature, the mountain lion in a city park provides hope for a new breed of relationship with nature, not the hands-off, take-care-not-to-anthropomorphize, us-verses-them way that scientists have preached for so long. This doesn't mean approaching P-22 and giving him a friendly pat. But it does mean seeing wildlife as part of the landscape, as part of our neighborhoods.

Wildlife isn't just about idyllic nature settings, or science or environmentalism, it's about art and culture and history and spirituality. In Los Angeles, wildlife is about coexistence, about human and nonhuman residents sharing space and adapting to life together in this grand metropolis. It is a coexistence that is fraught with difficulty, and that doesn't always have a happy ending,

New kittens P-46 and P-47 in the Santa Monica Mountains in 2015.

especially for the wildlife (don't fret—just wait for the sequel), but that ultimately can be beneficial to all.

"Nature has had a mixed career in Los Angeles," notes professor of urban studies Roger Keil, and I agree. But signs point to the city fast-tracking Mother Nature for a promotion. LAX, one of the largest airports in the world, makes way for the endangered El Segundo Blue butterfly by restoring habitat on its property, and Travis Longcore of the Urban Wildlands Group works throughout the region to expand these restoration efforts. Elementary school students at the UCLA Lab School, under the tutelage of watershed expert Mark Abramson, daylighted an entire creek on their campus and excitedly greeted the first black-bellied slender salamanders, a family of mallards that returns every year, butterflies, and a resident red-shouldered hawk. In Glendale, a whole community rallied around the famous black bear named Meatball, even fundraising for a place to relocate him after he stole one too many Costco meatballs from area homes (see page 33). Leo Politi Elementary School, in the middle of Los Angeles, installed an onsite wildlife habitat with the help of LA Audubon and the United States Fish and Wildlife Service. They added *The Sibley Guide to Birds* to their curriculum, and the students participate in the annual Christ-mas bird count (see page 193). Perhaps most tellingly, Mayor Eric Garcetti has moved forward a billion-dollar revitalization and res-toration project that will transform miles of the Los Angeles River. Progress is happening all around us, and an urban mountain lion is the ultimate sign of the region's ecological health.

The plight of P-22 has captured the imagination of Angeli-nos—and people across the globe—bringing them a glimpse into a wilder world, one that refuses to be contained by the boundar-ies of endless paved freeways. Even some who fear P-22 and his brethren still cheer him on, sharing the sentiment expressed by Gregory Rodriguez in the *LA Times*: "I have no illusions that the Glendale bear or P-22 wouldn't hesitate to dine on me given the right circumstances. But I'm still rooting for them. Deep down

I'm hoping that if they can survive at the margins of human civilization without forsaking their wildness, so can I."

Truly, it's something to celebrate that the city that gave us Carmageddon also has allowed a mountain lion to thrive. Los Angeles now needs to prove to P-22 his journey wasn't for naught. Let's give him—and all his Santa Monica Mountain kin—a Hollywood ending by building the largest wildlife crossing in the world in one of the largest urban areas in the country. He deserves as much.

Steve Winter: Photographing P-22

Steve Winter, an award-winning international wildlife photographer for *National Geographic* who has been "trapped in quicksand in the world's largest tiger reserve in Myanmar and slept in a tent for six months at forty below zero tracking snow leopards," says getting the famous photo of P-22 in front of the Hollywood sign was the most challenging thing he has ever done. His photos accompanied the story "Ghost Cats," which appeared in the December 2013 issue of *National Geographic*.

Using a series of sophisticated remote cameras and working with biologist Jeff Sikich, he tracked P-22 for fourteen months before capturing the remarkable image. "I am an eternal optimist," he says.

But Steve isn't just interested in a good photo. He is also passionate about protecting big cats worldwide, and he lectures globally on photography and conservation issues. *National Geographic* published Steve's recent photography book, *Tigers Forever: Saving the World's Most Endangered Big Cat* (with prose by author Sharon Guynup), and premiered his one-hour film, *Mission Critical: Leopards at the Door*, worldwide on its channel in January of 2016.

Steve hopes the photos of P-22 will help people appreciate these cats and show that we can live among cougars without many problems. "I also hope these photos will rally Los Angeles around building the Liberty Canyon crossing for the mountain lions," he says. "Living with these cats is something to celebrate."

National Geographic photographer Steve Winter.

Rat Poison: A Threat to P-22 and All Wildlife

In April of 2014, scientists released a startling photo of P-22 looking nothing like the majestic and handsome cat featured in *National Geographic*. He appeared weak and disoriented, ears drooping, whiskers bent, and his coat displayed red blotches of hair loss. He was suffering from a bad case of mange, caused by parasitic mites, probably due to his weakened condition as a result of having been exposed to anticoagulant rodenticide, commonly known as rat poison. A study by the National Park Service has documented three mountain lion deaths to date as a result of rodenticide poisoning.

"Anticoagulant rodenticides are designed to kill rodents by thinning the blood and preventing clotting," says Seth Riley, an urban wildlife expert at the Santa Monica Mountains National Recreation Area. "When people put these bait traps outside their homes or businesses, they may not realize that the poison works its way up the food chain, becoming more lethal as the dose accumulates in larger animals." Fortunately, biologists administered a treatment to P-22, and photos taken six months later revealed that he had recovered. But the threat remains—he could easily suffer the same consequences again by eating prey that has ingested the common poison.

Many groups are working toward eliminating this danger for P-22 and all wildlife, with efforts that include promoting natural solutions to pest control, such as building homes for barn owls to help with rodent problems. California has introduced a series of progressive legislation calling for bans of these substances, such as Assemblyman Richard Bloom's AB 2657, which took effect in 2015, and AB 2596, introduced by Bloom and state senator Fran Pavley, in 2016. Community groups like CLAW (Citizens for Los Angeles Wildlife) and Poison Free Malibu work to support lawmakers and educate people on how they can help. CLAW's executive director, Alison Simard, wants to implement more extensive regulations. "We want to increase the bans, in the interest of safety for people and wildlife. Poison is poison. If your pest control company says their stuff isn't toxic, ask them to lick it," she says.

P-22 suffering from mange.

Christopher Stills @chrisstills · Nov 29
pic.twitter.com/PbD8qyXSR

How David Crosby Advanced Mountain Lion Research

Another important discovery for the National Park Service's mountain lion study in the Santa Monica Mountains came from a citizen scientist sending in a photograph from a security camera. Since the lion in the picture doesn't have a GPS identification collar, researchers have no way of determining where it journeyed from, or even if it's male or female, but based on the location of the sighting, some theorize another mountain lion may have crossed the 405.

How did this discovery come to light? You can thank David Crosby and his friend Chris Stills, the son of Stephen Stills of Crosby, Stills, and Nash. David is a friend of mine, and I've kept him posted about the adventures of P-22 (along with pikas, frogs, and other wildlife). He alerted me one night in November of 2014 that Chris had tweeted a photo of a mountain lion, asking me if this was one being studied. I forwarded the photo to the National Park Service researchers and others, who noted this might be an unusual discovery.

It certainly adds some fun to citizen science efforts when it involves two great musicians. As KPCC radio's Jed Kim reported, "It's the kind of wildlife story that can only be told in Los Angeles, complete with a blurry, paparazzi-esque photo and celebrities of both the human and animal variety."

Mystery mountain lion roaming in Los Angeles.

Meet the Other Cougars of Los Angeles

P-22 is LA's most famous lion, but National Park Service researchers have marked forty-seven—and counting—in their ongoing study. Let's meet a few.

P-3: One of the first of ten lions to have been captured north of the 101, P-3 sadly died of anticoagulant rat poison in 2005.

P-12: An adventurer like P-22, P-12 is the first and only mountain lion known to have crossed the 101 from the north. He's fathered at least six litters of kittens, two with his daughter, P-19, in the fourth known example of inbreeding in this cat population. He fathered another litter with his granddaughter, P-23.

P-18: Captured and fitted with a tracking device as a one-month-old kitten, P-18 tried to find new territory as young adult but was unfortunately killed crossing the 405 near the Getty Center in 2011.

P-23: A young female—and proposed soul mate for P-22—P-23 made one of her first solo kills on the side of the busy Mulholland Freeway. The aftermath was captured in a photo by Irv Nilsen that included cyclist Danny Benson, who received the surprise of his life as he pedaled toward the lion enjoying her meal.

P-38: This male cat was captured in March of 2015 on the eastern end of the Santa Susana Mountains, north of Los Angeles. Researchers hope he will help shed some light on the impact of Highway 126 on carnivores.

P-45: Could this be P-12's challenger? This large 150-pound male surprised researchers by appearing on the scene in November of 2015. "During the course of our study, we've only been aware of one or two adult males at any given time in the Santa Monica Mountains. We're very interested to learn whether there are now three adult males or whether P-45 successfully challenged one of his competitors," said NPS biologist Jeff Sikich.

P-38 makes his home in the Santa Susana Mountains.

Making Cougarmagic: Tracking Mountain Lions with Cameras

For Johanna Turner, capturing wildlife via camera traps started as a fun hobby. By day she's a sound-effects editor for Universal Studios, but in her free time she wanders in the mountains around Los Angeles positioning and checking her photographic equipment, and then sharing her images on the site www.cougarmagic.com. "I started setting up cameras because I wanted to see what happened on some of my favorite hikes when I wasn't there. And it turns out a lot is happening," she says. Wildlife who have made appearances on her cameras include bears, bobcats, mountain lions, and bighorn sheep.

What began as a fun pastime turned into an important science project when Johanna captured a photograph in 2010 of a mountain lion in the Verdugo Mountains, which surround parts of Burbank, Los Angeles, and Glendale. Although larger than P-22's Griffith Park home, the Verdugos still offer a much smaller territory than is typical for a cougar. "Finding a cat there was 100 percent a surprise," she says. She and Denis Callet, another photographer, shared their images with National Park Service researchers, and Jeff Sikich captured and collared the animal in 2015. He's now known as P-41.

Johanna has relished joining the ranks of citizen scientists. "It's the best. I've learned so much from Jeff—it's important to me to do things in a way that is correct and worthwhile," she says. "I never thought of myself as a scientist. I was always bad at math. So to be helpful is amazing." She hopes that, aside from their scientific use, these images will enable people to better understand and appreciate mountain lions. "How can you not like these big cats?" she says. "They are beautiful, they are curious, they are not what anyone seems to think they are, and having them on the landscape means a chance to discover something."

Citizen science revealed P-41's presence in the Verdugo Mountains.

Meatball, the Glendale Bear

The Glendale Bear, affectionately known as Meatball for his successful raids of area homes in search of Costco meatballs, became famous for his neighborhood break-ins, and for the image a live news helicopter captured of the bear startling an unsuspecting resident while he texted on his phone.

Meatball also had his own Twitter account and conversed there regularly with P-22 about the challenges of urban living. Ultimately, because of his affinity for human food and relaxing in backyard hot tubs, this smarter-than-average black bear had to be captured in 2012 and sent to the animal sanctuary Lions, Tigers, and Bears in San Diego County, where he now resides.

The story, however, has a happy ending, as the residents of Glendale, aware their habits of leaving out trash and pet food might have made them culpable in his fate, promised to mend their ways and even raised funds for a bigger enclosure for Meatball. For the 2014 Rose Bowl Parade, Glendale created a float themed "Let's Be Neighbors," featuring Meatball in his famous trash-can pose. As Patricia Betancourt from the City of Glendale office said, "Glendale citizens, because of Meatball's influence, are now dedicated to being good neighbors to wildlife."

Meatball cooling off in a tub at his new home.

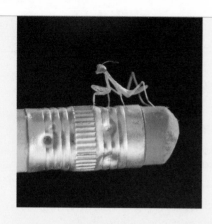

Charismatic Microfauna: The Bugs of LA

Lila Higgins, a self-proclaimed nature geek and urban nature explorer, has a mission: giving bugs their rightful due in the wildlife kingdom. Tired of the countless cuddly puppies and bunnies featured on the popular website Cute Overload, she knew she "had to get an insect on there," and ultimately succeeded with a photo of a tiny baby praying mantis. Her article for the *Smithsonian,* "The Everyday Cannibals and Murderers of Los Angeles," took a different tack, showcasing some truly scary critters lurking among us, like dragonfly nymphs that furtively swim in the LA River, using their jaws of death to capture tadpoles and small fish.

During a panel presentation I did in 2013 with Lila, the moderator asked us to nominate an "official" animal for Los Angeles. Not surprisingly, I voted for the mountain lion, while Lila offered the harvester ant, a highly social insect that forms its own cities and collaborative societies. "P-22 is charismatic," she said, "but if you look under a microscope, you'll discover a whole world of charismatic microfauna. Bugs help pollinate the food we eat and are a huge part of the food chain. You want big animals like bears? Well, you need grubs for them to survive."

As part of her work with the Natural History Museum of Los Angeles County, Lila also helps with the first-of-its-kind BioSCAN citizen science project, led by Brian Brown, which documents the diversity of insect species living in the city. A paper by Emily Hartop showed that, to date, they've discovered thirty new insects not previously known to science, and a host not known previously to reside in Los Angeles, including Lila's favorites, the phorid flies known as "ant-decapitating flies." Bugs are all around, even in a city, and Lila urges people to start taking notice: "The other day on Metrolink, I saw a fig wasp crawling on my jacket. How cool—even the bugs use public transportation to move around."

A baby praying mantis.

Calling in the Marines for the Desert Tortoise

In the battle against marauding ravens, increasing development, climate change, and a host of other challenges that has led California's official state reptile to be listed as a threatened species, the desert tortoise has acquired a formidable ally: the United States Marines Corps.

In the past thirty years, desert tortoise populations have plummeted by 90 percent as the surge of development has taken its toll. One result of this increase in human activity is the related population explosion of ravens, one of the biggest threats to desert tortoises. These intelligent birds know that where humans go, food and shelter abound, and they have accompanied our invasion of the desert. As a result, their populations have increased 1,000 percent over the last three decades, and desert tortoise hatchlings have little defense against this growing army. To a raven, the hatchlings, whose shells can take nine years to harden completely, look like walking ravioli, as one biologist described it.

The Twentynine Palms marine corps base partnered with UCLA to create the Desert Tortoise Head Start Program on a five-acre site that hosts five hundred hatchlings secured from predators and human activity until their shells harden. In 2015, the first thirty-five young tortoises in the program were released in the wild. Every marine on the base gets briefed on their duty to help the desert tortoise, from minimizing garbage that attracts ravens to halting training exercises if a tortoise is spotted in the area. "Along with military skills, we've also trained thirty thousand marines every year on desert tortoise conservation. Some of them pursue careers in wildlife biology as a result," says Brian Henen, director of the tortoise efforts on the base. One general even adopted two pet tortoises who could not be released to the wild; he, along with some boy scouts, named them Thelma and Louise, and after he deployed to Iraq they now serve as educational ambassadors for the facility.

Col. James F. Harp releasing a desert tortoise.

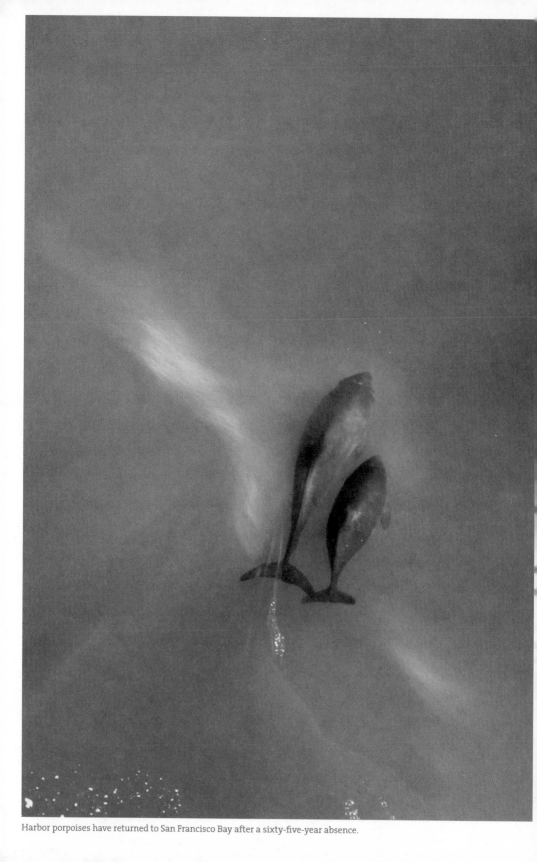

Harbor porpoises have returned to San Francisco Bay after a sixty-five-year absence.

CHAPTER 2

They Left Their Hearts in San Francisco

THE AMAZING RETURN OF THE HARBOR PORPOISE

"Compared to what is known about the coastal Common Bottlenose Dolphins and Killer Whales, very little is known about the life of the Harbor Porpoise. Harbor Porpoises are shy and tend to stay away from vessels, making them relatively inconspicuous, and difficult to study intensively."
—*Field Guide to Marine Mammals of the Pacific Coast* (2011)

"It would mean so much for everyone if the idea of restoration would catch fire with the public—the natural world is something to be recovered rather than just protected in pathetic remnants."
—biologist Edward O. Wilson, speaking to California State Parks

Four hundred miles north and five years before a mountain lion was discovered in the heart of Los Angeles, San Francisco Bay Area residents celebrated the homecoming of another shy animal, the harbor porpoise, who after a sixty-five-year absence returned to frolic in the bay. As has too often been the case with this lesser-known species, at first they were mistaken for their more charismatic cousins.

"Dolphins!"

The exclamation quickly gathered a crowd, all hoping for a glimpse of the creatures.

"See? There they are—near the concrete base," said a young man as he gestured to guide his girlfriend's eyes. Next to them a father held his child up for a better look, albeit cautiously, for the animals swam a dizzying 220 feet below.

Below? Where can onlookers marvel over marine mammals from such heights? In this case, they were observing the "dolphins" from one of the most recognizable landmarks in the world: the Golden Gate Bridge.

The bridge provides an unparalleled wildlife viewing platform; standing on the shoulder of a giant, you can watch marine life in its proper perspective, enveloped in the enormity of the ocean. Below the bridge and above the undulating greens and blues and browns and reds of the water's surface (painted by the light interacting with dispersing sediments), pelicans fly, surveying the ocean for a meal, and harbor seals propel through the tides or rest on rocky outcroppings. During their annual migration, gray whales, larger than some of the boats they pass, may even glide under the bridge. Viewers might see a seagull settling over the young in the nest it built on a concrete pillar, or a cormorant may suddenly break the water's surface—returned from his plunge for a herring snack.

And dolphins cavort in the waves. Sightings of bottlenose dolphins in all of California's waters tend to be hit or miss. Their population is estimated at only five hundred animals living from the Sonoma coast southward, and in the last few hundred years San Francisco Bay has been beyond the northern end of their normal range. After an El Niño event in the 1980s warmed the waters, however, they began heading north, following their food supply, and by 2007 they were making irregular visits to the bay.

When I heard the shout of "Dolphin!" I knew the specialness of this encounter. I hurried to the edge and leaned over, marveling at the smooth, dark-gray backs emerging from the silted waters, the animals leaping in their neat penmanship of question marks written by the ocean. Natural synchronized swimmers, marine mammals always seem exceedingly well choreographed. I watched those shapes dance again and again and stole glances at the crowd. Watching people watch wildlife can be as entertaining as watching the wildlife themselves. Delight infused every face. More and more people stopped to get a glimpse of the animals, gesturing excitedly to each other and snapping photos. Those stunning panoramic views of San Francisco and the Pacific coast were put on hold as the crowds admired the creatures that appeared toy-like in size from hundreds of feet above.

As the gusty wind whipped my hair, the animals disappeared under the bridge and into the Pacific. The mystery of the ocean had revealed itself that day and I, along with the crowd, went home satisfied that I had seen something remarkable: wild dolphins dancing under the Golden Gate Bridge. The photos I took of dorsal fins rising from the waves, although grainy even with my zoom lens, made for a popular Facebook post.

Except they weren't dolphins.

About a year later I listened to a story on NPR about harbor porpoises returning to San Francisco Bay after a sixty-five-year absence and how a team of scientists hoped to unravel the mystery of their reappearance. I immediately thought of my sighting that day on the bridge. I contacted researcher Bill Keener and sent him the photos I had taken. His enthusiasm and excitement was apparent in his return email: "These are great photos. Those are definitely porpoises. And it looks like they are engaging in mating behavior—it had never before been observed in the wild until our study."

Thus began my interest in the incredible story of the return of the porpoises to San Francisco Bay, and my foray into the world of porpoise porn.

I asked a pair of porpoises
what the purpose of a porpoise is—
and whether dolphins are like porpoises
to all intents and purposes?

"How dare you!" said the porpoises,
storming off to sea.
It was strange to have cross porpoises
talking at me.
—"Porpoises," Nick Asbury

Any discussion about porpoises needs to begin with a list of the ways in which they are different from dolphins, since most people don't distinguish between the two species. Unless you make your living as a cetacean scientist, you probably use the terms "dolphin" and "porpoise" interchangeably. Hemingway, in his *Old Man and the Sea,* related a dream of a "vast school of porpoises that stretched for eight or ten miles," the Grateful Dead sang of "mermaids on porpoises," and I am quite sure Mark Twain referred to dolphins, not porpoises, when he wrote, "You talk about happy creatures—did you ever notice a porpoise?—well there ain't anything in heaven here superior to that happiness."

Proof of this mistaken identity? Porpoises lack the grinning beak and joyous, playful manner of dolphins, they rarely travel in groups larger than a dozen, and they are notoriously shy of people (and, one could safely infer, also of mermaids).

Jim Carrey, in the goofy movie *Ace Ventura: Pet Detective,* chides his assistant for referring to the missing Miami Dolphins mascot as a porpoise:

Harbor porpoises are smaller than their dolphin cousins.

You see, nobody's missing a porpoise. It's a dolphin that's been taken. The common harbor porpoise has an abrupt snout, pointed teeth, and a triangular thoracic fin, while the bottlenosed dolphin, or *Tursiops truncatus,* has an elongated beak, round, cone-shaped teeth, and a serrated dorsal appendage. But I'm sure you already knew that.

Add to his description that the harbor porpoise has a darker coloration and is smaller in size—at an average five feet in length, it ranks as the smallest cetacean on the West Coast—and that's a pretty accurate inventory of the physical differences between the two animals. It also represents an almost complete inventory of what we knew, until very recently, about the harbor porpoise as a living being. Although even a layperson could probably rattle off interesting facts about dolphins just from osmosis of popular culture, the character of harbor porpoises has long eluded even researchers. In 2011, the *Field Guide to Marine Mammals of the*

Pacific Coast simply stated, "Very little is known about the life of the Harbor Porpoise."

These shy, introverted creatures seemed destined to remain one of the mysteries of the sea—until they decided after sixty-five years to venture into San Francisco Bay again in 2007. You can't remain elusive surrounded by seven million people. At just a mile wide, the narrow entrance of the San Francisco Bay acts as a small funnel that naturally herds marine life as they travel back and forth from bay to ocean, and the platform of the Golden Gate Bridge makes watching the antics of these animals especially easy. The stars have aligned in the bay to give the researchers unprecedented access, and in just a few years, Bill and his team at Golden Gate Cetacean Research—the nonprofit he formed to study the animals—have advanced the realm of harbor porpoise scholarship so significantly that the next edition of the above-mentioned field guide will have to amend its entry. San Francisco Bay is now the best place to study harbor porpoises in the world. As marine mammal scientist Izzy Szczepaniak boasts, "I've gotten more photos of porpoises in one day in the bay than in thirty years of study elsewhere."

Porpoises were native to the San Francisco Bay until they abandoned their homeland in the 1940s, driven out by, among other factors, the pollution that surged through the bay back then. Even when harbor porpoises are abundant, however, they remain elusive and are therefore hard to study. More than 9,000 inhabit the California coast around the San Francisco Bay Area, and an estimated 670,000 exist worldwide, but they are far less social than the bottlenose dolphins, who are easier to study in the open ocean as they swim frequently alongside boats and people. Harbor porpoises, by contrast, typically shy away from people, and they don't "cartwheel in the waves" like dolphins or vocalize in the human-audible chirps and whistles and clicks that have endeared us to dolphins. Indeed, porpoises hardly break the surface when coming up to breathe, they communicate in ultrasonic clicks beyond our range of hearing (and wisely beyond that of killer whales), and their forceful way of breathing in "chuffs" inspired sailors to give them the nickname "puffing pigs."

I can't find many accounts of a porpoise heroically saving a human from a shark or leading a person lost at sea back to land, and neither has anyone tried to make a go at a "swimming with porpoises" theme park. These creatures appear content to leave the human fraternizing to their dolphin and whale relatives—a characteristic that might support the case that they are of superior intelligence, especially given our mixed track record of protecting ocean life.

Nonetheless, we must use what intelligence we have to fill the gap in porpoise knowledge, because knowing how an animal behaves ensures we can help provide what it needs for survival. And that's why Bill's work is so important.

A better porpoise ambassador never existed than Bill. At our first in-person meeting, he donned his trademark porpoise-themed baseball hat, smiled infectiously, and whipped out a worn binder filled with a multitude of porpoise photos. He flipped through the catalog while describing the history of each animal in the animated voice of a natural storyteller, and he didn't even hesitate at showing the small group that accompanied me—which included National Wildlife Federation board members and some of their teenaged children—the photos of a porpoise in flight brandishing his eighteen-inch penis. "The penis accounts for almost a third of its body size," he noted unfazed, a fact I heard the teenagers excitedly reporting to others at dinner later that evening. He had successfully converted a few more porpoise advocates to the cause.

Bill and his colleagues at Golden Gate Cetacean Research (GGCR)—Izzy Szczepaniak, Jonathan Stern, and Marc Webber—have been looking at marine mammals collectively longer than anyone in the Bay Area. What started out as a group of friends collaborating on a nonprofit by volunteering countless hours and contributing their own funds to the cause eventually led them

Bill Keener photographing porpoises from the Golden Gate Bridge.

to form GGCR to "add to the body of scientific knowledge of the species inhabiting San Francisco Bay Area waters, and provide resource managers with the information they need to make wise conservation choices." Key research areas include the bottlenose dolphin, the minke whale, and the harbor porpoise. When not out on the ocean, Marc teaches and writes. He just finished an update of the definitive work he coauthored, *Marine Mammals of the World,* and he has also studied dusky dolphins, Pacific walruses, and northern fur seals. Izzy, an avid surfer who calls his work "an avocation rather than a vocation," has studied harbor porpoises for thirty-five years, and he has also worked in what's commonly referred to as "California's Galapagos," the Farallon Islands.

Rounding out the team is Jon, a marine biologist considered one of the world's foremost experts on minke whales. He got into whale research almost thirty years ago to "meet chicks," he tells me with the wry look of a college intern. Early in his career, he joined Kenneth Balcomb (a scientist featured in the documentary

Blackfish) as part of the team that has since compiled the famous photo catalog of orcas in the San Juan Islands, but in a sudden departure he ultimately chose to study minke whales instead. Although he loved the killer whale project, minkes were far less known, and he reasoned that if he studied them he'd "get to see killer whales while doing it anyway." Jon equates the minke to the harbor porpoise in terms of behavior and reputation: "They're tiny in the whale world and not as charismatic, and also very difficult to see."

It was Jon who made the landmark sighting of the first porpoise in the bay while out on his boat, doing his usual thing of scanning the waters for marine life. Much like the stunned researcher who stumbled on the first photo of Los Angeles's mountain lion, Jon couldn't believe what he was seeing at first. After marveling over the unmistakable blunt snout and almost square dorsal fin cutting through the waves, he called Bill to share the good news. Jon met Bill in the 1970s when he volunteered at the Marine Mammal Center, a model facility for pinniped rescue and research based in Marin County (see page 68), where Bill once served as executive director.

Bill wasn't sure whether to believe Jon's discovery at first, so he decided to verify it for himself. As Bill recounts, "The next day, I went out alone on the headlands near the Golden Gate Bridge and scanned the water with my binoculars. I was blown away. Sure enough, there were porpoises." The entire team was astonished by the reappearance of the porpoise. "All four of us were raised locally, and had our collectives eyes on the water since the mid-1970s," he said.

Bill now spends most of his free time searching for porpoises. Before this project, most of the limited body of work on harbor porpoises came from postmortem examinations of dead animals washed ashore, but dissection can only go so far in the study of animal behavior. Bill and his volunteer team—which has expanded to include graduate students attracted to this

The most easily recognizable porpoise in San Francisco Bay, a white animal nicknamed Mini-Moby.

first-of-its-kind work—have since exponentially advanced our scientific knowledge of the harbor porpoise.

In just a few years, GGCR has amassed the world's first photo library of harbor porpoises, logging six hundred individuals and counting. Using these photos, they monitor individual porpoises over time, comparing scarring and distinct skin coloration to make identifications. Gazing through their yearbooks, the researchers compare squiggles, oval spots, patches, and bumps to track information on specific animals. For instance, SFB 127—one of the oldest porpoises in the catalog, from the Class of 2010—has been seen every year and can be identified by the series of consistent parallel grooves on her left side. Unexplained white marks appear and disappear on her right side with each sighting. If it seems to you that pattern identification would be a time-consuming endeavor, then you are understanding the work these researchers do. The process requires insane attention to detail.

Sometimes nature cuts the researchers a break with a distinct-looking animal. Take Mini-Moby, an albino porpoise first spotted in the bay in 2011. "He's pretty hard to miss," says Bill. "He's definitely the most recognizable porpoise in the Bay Area, if not the entire coast of California." Almost entirely white, black accents decorate his dorsal fin and blowhole—the porpoise version of a tuxedo. Technically, this makes him "leucistic," not a true albino.

Another easily recognizable individual is Scoliosis. Bill and the GGCR team have spotted her nine times to date, as her unique body shape—a bit twisted, with a noticeable hump—along with the small wake she leaves while swimming, makes her easy to identify even from the height of the Golden Gate Bridge. Her condition may be due to a spinal disorder, or even a major injury, although an injury typically would have left scars and she displays none. No one expected her to survive very long.

In December of 2013, Scoliosis surprised the researchers by swimming past them not only healthy and alive but also

The hardy survivor, harbor porpoise Scoliosis, with her surprise calf.

accompanied by a calf. Bill expressed his admiration for the animal: "Because she is not 'normal,' we were not sure she could withstand the rigors of pregnancy, birth, and nursing, but her calf seemed frisky and doing well at about six months old. So we learned a lot about harbor porpoises just from this single observation, and it confirms that they can be tough little animals."

On their website, GGCR calls for citizen scientists—i.e., anyone with a camera—to submit photos of Mini-Moby, Scoliosis, or any porpoise they encounter in the bay (see page 62). Citizen science plays a large role in developing their photo catalog, and now it's easier than ever to get involved, since people can snap and upload photos directly on their cell phones; just a few clicks and you can help advance porpoise, dolphin, and other bay research. "If we can get a good enough [image] resolution, we can often identify animals," says Bill. "People have helped our effort enormously by sending us photos."

Citizen science programs have exploded in recent years, partially out of desperation resulting from shrinking research budgets, but the increased participation can also be attributed to a genuine desire of everyday people to connect deeper with wildlife and feel like they are making a difference for its future. We feel helpless when barraged by messages of polar bears drowning or elephants facing extinction, and these problems often appear insurmountable. Mary Ellen Hannibal, author of the upcoming book *Citizen Scientist,* observes, "The amazing thing about citizen science is that it has impact across scales. People do it because it is fun and gets you outside. It helps bring biodiversity lessons up close and personal. But it is much more than an outreach tool. As long as people use programs that link to scientifically sound databases, they are actually contributing to scientific research." Hannibal cites iNaturalist and Nature's Notebook as among her favorite citizen science platforms, available to anyone with a mobile device or computer. "To help us get ahold of the biodiversity crisis, we need massive amounts of data points," she continues. "Citizen science is a way to help find solutions to the greatest challenge we all face, which is the massive loss of species at a rate and magnitude that took out the dinosaurs. It is a hopeful way forward."

Citizen science offers a way for everyone to take an active role in conservation, and technology has made citizen science extremely accessible. People have signed up to monitor everything from bees in their backyards (Gretchen Lebuhn's Great Sunflower Project has more than one hundred thousand participants) to raptors flying over the bay (see page 67) to pikas on high mountain peaks (see page 105) and mountain lions in Los Angeles (see page 32).

Porpoises are no exception. Bill is amassing a small army of dedicated citizen scientists to his cause, ranging from spontaneous participants to organized groups. For instance, a random individual reported a sighting under the San Francisco–Oakland Bay Bridge, a place Bill had not known the animals to frequent. And ten-year-old Chloe Helmlinger, who found a stranded porpoise on a beach in Pacifica, sent a photo to Bill and did a class presentation on her encounter. Perhaps Bill's most dedicated citizen scientist is Bob Kaltreider, who has tracked the porpoises since their return and took the first known photo of a wild porpoise in the bay. He recently spent weeks searching for a porpoise that had ventured all the way up to the Napa River. After having no luck for several days, he decided to go home for lunch and make himself a tuna sandwich. While eating, he watched in astonishment as the porpoise swam right by the dock on his lagoon. He managed to snap a photo, which he quickly and proudly shared with Bill and the local newspaper.

Like these people, I don't just want to see wildlife, I want to participate in their preservation, to feel like I have contributed even a small part to their survival. We tend to think of wildlife in terms of herds or pods or schools and not as individuals. Tracking their lives erases those generalities and helps us consider them each as unique beings. This is the future of conservation—building relationships and connections. As animals ourselves, we can all

Bill Keener and his team have identified more than six hundred individual porpoises in their study.

identify with a wild creature's need to find a good home or keep a child safe, or even just indulge a preference for anchovies. These animals are more than abstractions in some scientific paper.

I've spent many a day standing on the Golden Gate Bridge, frozen from the relentless wind despite being huddled under layers of clothes. Glancing toward Russian Hill, I've been tempted by the promise of warm Irish coffees at the Buena Vista Cafe, but I never cut my vigils short. I kept thinking my next sighting could be important, another part of the story. Is SFB 127 still alive? Will Mini-Moby return another year? Will Scoliosis and her calf survive? I want to help tell these tales.

"On the entire west coast of the Americas, there is no other estuary like San Francisco Bay. Immense in size, covering over sixteen hundred square miles, and draining over 40 percent of the land area of California, it is one of the great estuaries of the world." —*Saving the Bay: The Story of San Francisco Bay,* a documentary film by Ron Blatman (2011)

A harbor porpoise pulses his strong flukes, the swirling currents no deterrent to this tough and fast swimmer. His clicks send through the murky water invisible ultrasonic waves, silent to humans and herring alike. Once he locates his quarry, he twists and turns in the water, pursuing the fish, using his small, spade-like teeth to catch a meal. Surfacing to breathe, he gives a trademark chuff and submerges again, but doesn't dive deep, as he prefers shallower water to the depths of the ocean. He must consume a tenth of his body weight daily, and the San Francisco Bay provides a rich source of herring and anchovies for him to nibble on.

The San Francisco Bay is a living, breathing entity, inhaling and exhaling enormous quantities of water. Seven times the water flow of the Mississippi Delta converges here, as snowmelt from the Sierra Nevada two hundred miles away unites with the Pacific Ocean, the largest body of water in the world. The tendency is to regard the bay as a tourist attraction, or an accumulation of ones, including the Golden Gate Bridge, Fisherman's Wharf, the Giants' baseball stadium, and the Ferry Building. Like Los Angeles's Griffith Park, the bay seems more an extension of the city than an ecosystem, merely a scenic vista to photograph or a place for recreational activities. People jog on the shoreline trails at Crissy Field, sail around Angel Island, or ride ferries to commute to work. It's a peopled landscape, not revealing at first the depth of wildlife it houses. Since we are taught to think of nature as existing only in pristine settings, a shipping barge the size of a house floating strongly across the bay waters suggests not-nature to us.

But if it's good enough for harbor porpoises, it's nature.

And harbor porpoises aren't the only animals giving the bay their stamp of approval. Despite being surrounded by seven million residents and ringed by an almost circular cityscape that hosts two major shipping ports and is traversed by eight bridges, this sixteen-hundred-mile estuary supports a diverse variety of life and boasts one of the most productive marine habitats on earth. More than 120 species of fish live in the bay, including Chinook salmon, steelhead trout, herring, and leopard sharks, which in turn provide the meals for California sea lions and harbor seals. A key part of the Pacific Flyway, the bay is home to 281 species of resident and migrating birds, among them the western sandpiper, the bufflehead, and the great blue heron. Peak migration can bring one million birds per day to the bay's shores.

It's a dynamic being, guided by the ebb and flow of the tides. On a boat you can't help but feel like you're in the middle of a stirring witch's cauldron; eddies and currents and flows boil up to the surface. And this storm of water unleashes on the bay floor as well, shaping and reshaping what lies beneath, creating massive, undulating sand dunes resembling the ripples on the underside of a dog's mouth. As Marie De Santis notes in her book *California Currents,* "Nowhere on the coast are the tidal vagaries more exaggerated and magnified than in San Francisco Bay."

The tides bring the plankton, the plankton bring the small fish, and the small fish bring harbor porpoises, and at regular times that make them easier for researchers to track. Porpoise-watching excursions are dictated by the tidal schedule, with an hour or two around high tide being the ideal viewing time. Since the porpoises have returned to the bay, sighting them is an almost never-miss endeavor in this narrow body of water that funnels them into view.

During one boat trip I took with Bill and his team, we weren't on the water fifteen minutes before the first porpoise was spotted, about a quarter of a mile offshore from Sausalito. As we sailed under the bridge, calls of "Thar she blows!" became frequent, for who can resist that declaration (even though porpoises snort rather than really blow)? The young daughters of two other guests on our boat, Esme and Elsa, ages eight and ten, settled at the bow and proved to be talented marine mammal spotters. The wind played with their hair and the sun reflected off the water to illuminate their faces as they laughed and jumped up and down with uncontained excitement whenever they spotted an animal. They were so captivated that they later mailed a thank-you letter and generously donated their allowance to help with porpoise research. That day we photographed twenty-six individual porpoises in our eleven-mile journey, and noted five with unique markings.

After our excursion, Bill invited me back to his home to view some new video footage. He had just spent time filming with a PBS crew for a segment in the series *Sex in the Wild,* for which he accompanied the show's host, Dr. Joy S. Reidenberg, on a quest to film porpoises mating in their natural habitat. They succeeded in obtaining the "first tantalizing glimpse of harbor porpoise sex ever recorded on film," according to the narrator's introduction. Just as the Golden Gate provides a narrowing point for viewing the animals eating, it does the same for the porpoise dating scene: "Jon jokingly refers to this as the funnel of love," Bill reveals.

Porpoises employ a rather dramatic and difficult hit-and-run approach to copulation. The male aims his large appendage, trying to penetrate the female, often jumping into the air as he rushes past her. All this lasts just one to two seconds. The maneuvering required for a successful mating encounter reminds me of the scene in *Star Wars* where Luke Skywalker has to hit a two-meter target while flying a fast-moving X-Wing fighter in order to destroy the Death Star; I am surprised porpoises didn't go extinct long ago. The more I learn about these little cetaceans, the more I am amazed at their hardiness.

There is something fascinating about a mating ritual so radically different than most mammals, yet the interest isn't just voyeuristic. Understanding the reproductive habits of these animals can not just help us gain new insights about their behavior overall but also help us better understand what conditions they need to survive. "We need these answers in order to ensure porpoises remain permanent residents," says Bill. "And part of the answer is tied up in the history of the San Francisco Bay itself and the changes humans brought to it over time."

Porpoise porn: harbor porpoises attempting their hit-and-run approach to mating.

> "With surprising rapidity, the movement to save the bay became a mass political uprising...a widely popular cause, and hundreds of people (including me) were converted to environmentalism in the process."—Richard Walker, *The Country in the City: The Greening of the San Francisco Bay Area*

Given the state of the bay today, it's difficult to imagine that locals had nicknamed it "The Big Stench" in the 1960s and literally held their noses when they approached the shoreline. As John Hart recounts in *San Francisco Bay: Portrait of an Estuary,* "In 1960, most of the bay's shoreline was closed to the public...often for good reason....'Anything that stank or was dangerous,' one observer notes, 'wound up on the bayshore.' That strictly utilitarian shoreline was a place of refineries, military bases, explosives factories, firing ranges, ports and airports, sewage outfalls, and dumps. It was not a place for enjoyment."

Of course, it wasn't always that way. Almost two hundred years earlier, in 1779, Juan Bautista de Anza, leader of the expedition of soldiers who first settled San Francisco, deemed San Francisco Bay, "a marvel of nature." Yet almost as soon as Europeans arrived, proponents of "progress" forcibly pushed nature aside, and the bay's long environmental decline commenced. During the gold rush, waste from mines far away in the Sierra foothills washed into rivers, which then daily transported tons of uncovered sediment into the bay. A California resource agency reported in 1879 on the "constant fouling of the waters and consequent destruction of life by the foetid inpourings of our sewers." Near the turn of that century, oil refineries appeared on the shores, adding yet another toxic ingredient to an already murky chemical soup. As the population continued to grow, so did the pumping of sewage, and by the 1970s, municipalities were adding 786 million gallons a day to bay waters.

San Francisco Bay had earned the unfortunate distinction of being the most altered estuary in the world, and the harbor porpoises had good reason to remain absent for so long.

Harbor porpoises dwell as far north as the Bering Sea and as far south as Central California. They live in both the Atlantic and the Pacific, but the two oceans house different subspecies that diverged genetically about five million years ago, and which came first is the harbor porpoise equivalent of the chicken and egg question. Unlike their counterparts that migrate along the Atlantic coast of the United States, Pacific porpoises tend to stay in smaller areas, clustered by subpopulations that keep to distinct regions along the coast from California to Alaska, and they live mostly within five to ten miles of shore.

Porpoises have a long history in California. Evidence of them in the Bay Area extends back almost twenty-six hundred years in the forms of bones tossed into shellmounds by local Native Americans. When archeologists later excavated these sites, they dated the porpoise bones from between 700 BC and 1300 AD. Most coastal tribes consumed porpoises, and the creature also had a place in their mythologies, including stories about the immortal "porpoise people" revered by the Yurok.

After Europeans arrived, porpoise sightings continued to be common in the bay. Captain Charles M. Scammon was a whaler turned naturalist who, according to author Dick Russell, "gave us our first enlightened look at whales." In Scammon's 1874 book *The Marine Mammals of the North-western Coast of North America,* he described what was then called the bay porpoise as a "peculiar species of dolphin" that "feed upon fish, and are occasionally taken in seines that are hauled along the shores of the San Francisco Bay by Italian fisherman." Fourth-generation ferryboat captain Maggie McDonogh recalled for *National Wildlife* magazine her father describing porpoises "snorting" in the local Tiburon harbor in the 1920s and 1930s, as does Ron Clausen, a Point Richmond business owner whose father recounted a childhood spent exploring the porpoise-filled bay.

Suddenly, in the 1940s, sightings ceased. Harbor porpoises had disappeared from the area.

Given the circumstances, it's clear why a porpoise cruising the coastline didn't have much incentive to make the turn under the Golden Gate. The 1940s was also the decade when the military placed a submarine net and underwater mines at the entrance to the bay, a move that may have been the last straw for the harbor porpoises. Swimming through chemicals and sewage was bad enough, but having to navigate an obstacle field of explosives proved too high a price for a herring snack. Fish were plentiful elsewhere, and probably lacked the petroleum marinade.

Porpoises weren't the only creatures that had given up on the bay—so had people. The dominant philosophy that prevailed for more than a hundred years was to banish nature rather than

A harbor porpoise near Golden Gate Bridge.

preserve it, and this had almost been accomplished by the 1960s, when development had reduced the open water of the bay by 50 percent and the surrounding wetlands by 80 percent. In 1962, however, the tide turned, both literally and figuratively. Bay Area residents owe the return of the harbor porpoise to the efforts of three Berkeley women who almost sixty years ago advanced the notion that cities are not devoid of nature, and that a landscape marred by humans isn't irredeemable. These women—Catherine Kerr, Sylvia McLaughlin, and Esther Gulick—became outraged after viewing a map showing the plan to fill in most of the remaining open water, essentially leaving a river where a massive bay had once been. As incredible as it sounds now, the idea of restoring something lost rather than preserving something untouched was considered at that time almost heretical.

These women also sparked something larger: the modern environmental movement based on localized and urban grassroots advocacy that influenced people worldwide. Prior to their efforts, urban conservation and restoration was simply not a priority. Nature was still something you journeyed to, an ethic solidified by John Muir's popular reverence for high, pristine places. The environmental powerhouses of the day also chose to focus on "traditional" wilderness, and indeed with David Brower, the Sierra Club, the Save the Redwoods League, and other groups claiming that preserving redwood forests and the Sierra Nevada took precedence over the bay, Kerr, McLaughlin, and Gulick were forced to form their own group.

Undeterred, they founded the Save the Bay Association (now simply Save the Bay) and rallied residents to join for a dollar each. Using the slogan "BAY OR RIVER?" their efforts ultimately rescued from destruction one of the most significant estuaries in the world. Environmental victories by others followed, including the Clean Water Act of 1972 and, that same year, the banning of toxic chemicals like DDT. Agencies also started to focus on wetland restoration rather than removal, and in 1974 the southern end of the bay became the site for the nation's first urban national wildlife preserve, the Don Edwards San Francisco Bay National Wildlife Refuge.

As the bay got cleaner, the wildlife slowly began returning on its own. Other critters took the initiative and decided to check out the new and improved digs as well. Sutro Sam, the first river otter seen in San Francisco in more than fifty years, appeared one day spinning playfully in the ruins of the Sutro Baths and became an instant celebrity (see page 63). Beavers began resettling in the area, most notably the population that built a dam in the middle of Martinez, located on the south side of the bay's Carquinez Strait (see page 65). Native fish, including the leopard shark, flourish again, benefiting from restored wetlands, many of which are the result of ongoing multiagency partnerships, such as the South Bay Salt Pond Restoration Project, whose scientists have already converted an industrial space the size of Manhattan back into natural habitat.

The return of the porpoise—and other wildlife—to San Francisco Bay serves as both a vindication of the Bay Area's past efforts and also a rallying cry for future ones. But the animal's continued presence in the bay is by no means assured. Bill's research team still needs to answer some key questions about habitat quality to pinpoint this creature's tolerance for bay living as water conditions fluctuate dramatically. The National Wildlife Federation has partnered with him to support his important work. Yet despite the amazing ongoing efforts of many environmental groups and government agencies dedicated to restoration and protection, this dynamic estuary still faces perhaps its most formidable challenges; the effects of climate change and drought are already becoming apparent.

But more importantly, people have to keep caring about the health of the bay—and its inhabitants. Captain Maggie, who loves taking people on tours of the bay, agrees. As she told Anne Bolen in an interview for *National Wildlife,* "I love seeing things come back around. I have three children that are growing up. I want them to be able to get out on the water and see the seals, the harbor porpoises, and the birds—and enjoy all of it. If people don't know about them, what has happened to them, they can't be inspired to care."

The Bay Area community appears up to the challenges. If Los Angeles's environmental consciousness, now emerging full force, was awakened by the unburying of a river and the presence of one lonely mountain lion, San Francisco's journey began more than sixty years ago with the outrage of three women. Today, porpoises swim in the bay largely as a result of their incredible efforts. So why not aim high for the future?

What else will return if we continue to roll out the welcome mat of restoration? Will sea otters float on their backs in the bay while prying open abalone? Will condors soar above the Golden Gate Bridge? There's a new movement afoot to reintroduce grizzly bears in California (see page 101); will someday they feed on the spawning salmon in Redwood Creek? Too outrageous? Don't dismiss it. This is San Francisco, after all, a city Paul Kantner of the band Jefferson Airplane called "forty-nine square miles surrounded by reality."

Admittedly, a grizzly might be out of reach, yet in the long-standing rivalry between LA and the Bay Area, I might remind you that Los Angeles has upped the ante with its mountain lion. Your move, San Francisco.

Become a Porpoise Citizen Scientist

Harbor porpoises live in the Bay Area's back-yard, and unlike many species of wildlife, you can observe them on a regular basis year-round. Golden Gate Cetacean Research scientist Bill Keener gives these tips for how to see them:

Cavallo Point, at Fort Baker in Marin, is a good place to observe porpoises at the beginning of the ebb tide, when the animals may be as close as thirty feet off the rocks. In San Francisco, porpoises can sometimes be seen working the flood tide off the pier at Crissy Field. A knot of gulls on the water often means a seal, sea lion, or porpoise is actively feeding there, so watch carefully. And bring binoculars.

But the best place to watch for porpoises is from the Golden Gate Bridge's pedestrian walkway. Check the tide tables and time your visit to within an hour or two around a big high tide. There are two good zones—one near the north tower and one near the south tower. Walking slowly all the way across the bridge and taking time to scan the water frequently will almost always result in a few sightings.

Here's your opportunity to become a citizen scientist. Information about your sightings of porpoises or dolphins is valuable, and researchers would love to have it. Add your photos to the GGCR's growing catalog by submitting your images at www.sfbayporpoises.org.

Harbor porpoises can easily be spotted in San Francisco Bay.

Sutro Sam: San Francisco's First River Otter in Fifty Years

In 2013 a river otter entered the spotlight when he suddenly appeared gliding and rolling playfully in the waters of the Sutro Baths, the ruins of a nineteenth-century public bathing pool on the edge of the Pacific coast in San Francisco. Visitors from all over California—and the world—traveled to get a glimpse of this celebrity otter after the *Huffington Post* and other notable media outlets featured his story. Why the fascination (aside from his being almost unbearably cute)? Because the return of the river otter after such a long absence underscored a conservation success story.

Megan Isadore, cofounder of the River Otter Ecology Project, celebrated his return for just this reason. "River otters are a sentinel species in that they require healthy watersheds to thrive," she said. "The fact that river otters can live all over the Bay Area indicates that we humans have done something right. It shows that we can make positive changes to our environment that allow wildlife to return and thrive in areas where we haven't seen them in a long time."

Scientists are still unsure of the reason for Sutro Sam's sudden appearance, but some think he might have traveled from nearby Marin, where the river otter population has been making a comeback. One attraction to Sutro Baths was certainly its unique and plentiful food source: for reasons unknown, people had been releasing pet goldfish into the baths for years, and this made for an easy meal for the hungry critter. Sutro Sam has since left the site, probably in quest for romance, yet he serves as a hopeful sign that river otters might someday become common in San Francisco waters once again.

Sutro Sam at Lands End, San Francisco.

Lake Merritt: Connecting People and Wildlife in Oakland

Another comeback kid on the roster of San Francisco Bay success stories, Lake Merritt is a 155-acre tidal lagoon in downtown Oakland that also has the distinction of being the first wildlife refuge in the country. Following years of decline, the lake's stagnant water and littered shores had once made it an eyesore, but then the community rallied and in 2002 local voters agreed to fund restoration efforts. Nature responded with a noticeable increase of wildlife, and in 2013 a river otter paid the ultimate compliment by hanging out on a shoreline dock. He's the first to be seen in the area in decades.

"That was an amazing sighting at the lake," says Rue Mapp, an Oakland native and the founder of the nature organization Outdoor Afro. "There probably isn't a day that goes by when I don't mention the lake," she says. "It's part of our social fabric." Rue sees the lake as a vital connection point for Oakland—one of the reasons residents feel so passionate about restoring it. "Almost everyone I know has a memory of feeding ducks stale bread crumbs at Lake Merritt," she says, before adding with a laugh, "that was before we knew the harm—you now go into the Nature Center to get birdseed."

Lake Merritt serves as a great example of why we need to expand our view of "nature"; feeding the ducks in a city park as a child can lead to a person becoming an environmental champion on a larger scale. As Rue observes, you can't help but have a connection with the bay living in its presence: "No matter where you live on the bay, you develop a relationship to that water, whether walking at lunch hour at Lake Merritt, fishing at Martin Luther King Regional Park, or simply driving over a bridge. We need to respect the diversity of those connections."

Rue Mapp kayaking on Lake Merritt with her son, Will.

A Community Rallies Around the Martinez Beavers

People line the small footbridge that crosses Alhambra Creek in downtown Martinez, eagerly scouting the water for a glimpse of the city's star attraction. One young girl starts pointing excitedly at the water and exclaims, "I see bubbles! I see bubbles!" Soon after, the telltale ripples appear on the surface, announcing the arrival of the creature we have all gathered to see.

"A beaver! There's the beaver!" yells a teenaged boy next to me.

Most of the "eager beaver" watchers were attending the annual Beaver Festival, hosted by the nonprofit Worth A Dam to celebrate one of nature's best engineers and to honor the way in which the Martinez community rallied around its local beavers, who were slated to be removed in 2007 after being deemed a flood hazard. To save the beavers, local citizen Heidi Perryman formed Worth A Dam and started working with a beaver biologist to develop solutions for peaceful coexistence, such as coating trees with "sand paint" to prevent them from being chewed on, and installing "beaver deceivers," which curb dam building in certain areas.

Beavers are pretty remarkable animals. Weighing up to seventy pounds, they are the largest living rodent in North America, and they are excellent swimmers who can remain underwater up for to fifteen minutes. In the animal world, their monumental feats of engineering are rivaled only by humans; as Alice Outwater notes in her book *Water: A Natural History:* "Beavers do more to shape their landscape than any other mammal except for human beings, and their ancestors were building dams ten million years ago."

After the festival, I joined the crowd again to watch a beaver family adding branches to its already impressive dam. Onlookers clapped and cheered. I appreciated the crowd, the ability of the beaver to inspire, and the willingness of Perryman and others in the city to make room for this industrious little animal.

Artist FROgard Butler and children at the Martinez Beaver Festival.

The Pollinator Posse to the Rescue of Monarch Butterflies

Thousands of visitors flock every October to Pacific Grove, a.k.a. "Butterfly City, USA," to view the spectacle of the migrating monarch butterflies. When Laura Tangley, a writer for *National Wildlife,* visited the site, she "found the air filled with butterflies—scores of bright, orange-and-black monarchs fluttering among the branches," as they gathered to fuel up for the rest of their journey, which can total twenty-five hundred miles.

Dedicated scientists and volunteers have tracked the butterflies here for almost two decades, and their latest data are alarming. California's monarch population has decreased by 80 percent from its high point, and this trend is not unique to the state. Populations have declined by as much as 90 percent across the United States. The problem? The plentiful "gas stations" of milkweed and other plants needed to sustain them along their migration route have largely disappeared.

But there is good news. Helping the monarchs recover is easy, says David Mizejewski, a naturalist with the National Wildlife Federation: "We can each be a part of saving the monarch butterfly. The simple act of planting milkweed with your family provides monarchs with a place to lay their eggs, and helps ensure this iconic species has a future." In 2015, the NWF partnered with the US Fish and Wildlife Service as part of a nationwide campaign to save the imperiled monarch by encouraging people across the country to make their communities butterfly friendly.

Oakland's Pollinator Posse, an all-volunteer group, has also rallied to the monarch's cause, working to create pollination corridors across the city. The posse member leading the charge is parks supervisor Tora Rocha, who helped transform the Gardens at Lake Merritt and Children's Fairyland into butterfly-friendly gardens. "We help people get started with creating homes for monarchs," she says. "It's exciting and also addictive to watch those caterpillars transform into butterflies."

A monarch butterfly at Pacific Grove.

On Urban Raptors: GGRO Director Allen Fish

The Golden Gate Raptor Observatory, with its hundreds of dedicated volunteers, studies raptor populations along the Pacific coast just north of San Francisco in the Marin Headlands, an area that ranks as the most productive hawk migration site in the state. Allen Fish has directed the organization since its beginnings in 1985.

How did you get involved in urban raptor research? When I arrived at the GGRO in 1985, I was thinking how sad it was that I wasn't going to be [writer and environmentalist] Farley Mowat in Canada, studying wolves in the remote tundra. Instead I was surrounded by a density of human activity. "Couldn't we move the migration to Mendocino or Yosemite?" I thought. I didn't get it for a few years—these meeting places between the wild and populous human areas are vital to study. No other raptor migration site is as surrounded by as urban an area as we are, so if we're going to learn about how birds fare in the city environment, this is the place.

What makes the area such an important migration site? We are in this fantastic landscape with the San Francisco Bay in the middle of two symmetrical peninsulas—Marin County and San Francisco—that provide two launching sites that condense raptor movement. From a research perspective, it's been really exciting to see these birds survive in human environments.

What are some surprises from your research? We recently tracked a young red-tailed hawk that broke all the rules of migration. Instead of flying south from San Francisco, she circumnavigated the entire Sierra Nevada crest and even part of the Cascades. Why she did that we have no clue. But with GPS technology getting better and smaller, we can start to unravel some of these mysteries.

Broad-winged hawks on migration over the Golden Gate Bridge.

Saving California Sea Lions at the Marine Mammal Center

Heartbreaking photos of starving sea lion pups in California appeared all too frequently in news outlets across the world in 2015, and two nicknamed animals became the poster children for the recent plight of the state's sea lions. There was Rubbish, photographed resting his weary head on the curb of a sidewalk on a busy street, and Percevero, being rescued by a park ranger after he somehow managed to cross a four-lane road. Both had wandered onto the streets of San Francisco in a desperate search of food.

They were just two of thousands of sea lion pups stranded on California beaches in 2015, part of an alarming trend that began in 2013 with record-breaking numbers of pups turning up on California shores. The culprit is suspected to be warmer waters, which impact food availability for nursing mothers and newly weaned pups.

Whatever the cause, thousands of starving and stressed sea lions need care every year, and the Marine Mammal Center in Sausalito responds to many of these cases, as it did when it rescued Rubbish and Percevero. In 1975 the founders of the center started with "little more than kiddie pools, garden hoses, and ambitious dreams," and the complex now ranks as the world's largest rehabilitation facility, having rescued and treated more than twenty thousand marine mammals from more than six hundred miles of California coastline. In 2009, the center finished a $32 million rebuild of its headquarters—a combination research center, education facility, and the first purpose-built marine mammal hospital in the nation.

But despite the impressive new facility, the bricks and mortar of its existence continue to be its dedicated cadre of more than a thousand volunteers who enable it to respond to so many rescue cases. "The Marine Mammal Center exists because the community wants us here," says executive director Jeff Boehm. "Our story is one of collaboration and community support. More than twelve thousand volunteers support the fifty staff in all areas of operations, from animal care to administration. And the most rewarding part for the community is to watch hundreds of these rehabilitated animals released back to their ocean homes each year."

The Urban Coyote

Perhaps no large mammal has better adapted to a peopled landscape than the coyote. Scientist Stanley Gehrt "couldn't find an area in Chicago where there weren't coyotes," and one lone animal was even spotted on a rooftop in New York City; in Los Angeles, coyotes are a fact of life. Once revered by some Native American tribes, today the coyote often invokes fear, especially when featured in headlines that focus on rare attacks of people or pets. A number of advocates for coexisting with *Canis latrans,* however, are educating the public about this misunderstood creature—and the benefits of sharing our space.

Camilla Fox, the executive director for Project Coyote, offers innovative programs like Coyote Friendly Communities, which provides urban and rural audiences with the resources they need to foster safe coexistence with their wild neighbors. Her group worked recently with the City of Calabasas, which was spending $30,000 a year on coyote removal. As a result of their recent efforts, which have included demonstrating how the coyote helps with rodent control, the city council passed a resolution prohibiting expenditures on trapping and instead now works on educating residents about coexistence strategies.

In 2015, Jaymi Heimbuch launched "The Natural History of the Urban Coyote." Created alongside fellow photographers Morgan Heim and Karine Aigner, the photojournalism project brings together science about coyotes with intimate portraits of the species. "I want my photos to provide a window into the life of coyotes so people can understand them and not feel threatened or scared," she says. A resident of San Francisco, Jaymi began taking an interest in the coyote about a decade ago, when sightings in the city started increasing. "They are an integral part of nature," she notes. "When they returned to the Presidio, quail and other native birds began rebounding as a result of them helping to control fox and feral cat populations. The ecology in the park is completely different now that the coyotes are back."

This page: A coyote strolling the streets of San Francisco.
Opposite page: Rubbish takes a nap on a San Francisco sidewalk.

An estimated twenty-five to thirty thousand black bears live in California.

CHAPTER 3

Keeping Bears Wild

HOW STAFF AND VISITORS IN
YOSEMITE NATIONAL PARK HELP WILDLIFE

"Seeing a wild bear in its natural habitat is one of the most exhilarating aspects of my job. I'm over the moon when I see wild bears doing wild things."
—biologist Ryan Leahy, National Park Service

"Of all the animals in our national parks, the bear undoubtedly ranks first in public interest. There is something so human in the animal that its droll antics are doubly appealing to man. Even its apparent laziness is amusing, and the playfulness of bear cubs is a never-ending source of delight."
—M. E. Beatty, *Bears of Yosemite* (1943)

Here are forty-eight hours in the life of a Yosemite Valley black bear, inspired by the actual travels of a park research bear in August 2014:

6 p.m., Sunday: You wake up after a nice daytime nap near the oddly named but very popular Housekeeping Camp and begin sauntering around its structures, hybrid mixes of cabins and tents sheltering hundreds of visitors a day. People have returned from a day of sightseeing and are starting dinner preparations by lighting campfires or heating up gas grills. The aroma of steak, chicken, and fish being grilled, of salsa and guacamole being dipped with corn chips, and of marshmallows and chocolate being melted on graham crackers assaults you, as it does every night, this enticing bouquet as familiar as the fragrance of the surrounding trees and granite. You have been called a "nose on four legs" and possess one of the keenest senses of smell of any animal on earth, so the savory aroma of this human cornucopia is inescapable, even miles away from its source.

As you wander near the camp, you stay out of sight and go unnoticed by people as you surreptitiously scan the area, continually vigilant for unwatched coolers or containers of food carelessly left on picnic tables. You make a pass around the bear lockers—the secure metal units in which campers are required to store their food—sniffing them to detect what is inside, and pulling on the latches hoping someone forgot to lock one of the doors. The dinner smells become more abundant and irresistible as the evening progresses, and you patrol the area one more time, wondering if anyone absentmindedly abandoned his hot dogs for a minute or two to use the restroom. No luck. Campers here got the message, delivered when they registered, about the importance of food-storage regulations. There is ample wild food, but it takes more effort to obtain than these sometimes easily accessible human foodstuffs.

10 p.m., Sunday: The night is still young and you're not yet willing to give up on the promise of human carelessness—it's delivered many a meal before—so you decide to check another popular visitor area. You head northwest, roaming through the now deserted Yosemite Village. You can smell the lingering odor of the thousands of cold-cut sandwiches served at Degnan's Deli and the thousands of burgers and fries ordered at the Village Grill earlier today, and you wonder if a forgetful park visitor abandoned some leftovers on a picnic table or neglected to secure the bear-proof trash containers after disposing of his waste.

Nope.

You're foiled again, as the diligent park staff have made sure the trash cans near the fast-food places have been emptied and secured prior to your evening patrol, and they picked up any improperly discarded garbage as well. Not even a random ice-cream wrapper or lone French fry, you lament. Behind one building, you check out the larger dumpster and smell the food waste inside it, but you know it's pointless. Despite your impressive dexterity and ability to open doors and latches, the simple clip locking the hatch defeats your bulky paws.

You travel south now and try the Yosemite Lodge parking lot, hoping one of the visitors will have proven absent-minded and left a grocery bag of food in her car, perhaps a partially eaten bag of cookies, or maybe even just a stick of gum. You peek through a few windows and sniff the air—you can smell food in an automobile up to three miles away—and survey the parking lot. With your impressive strength, breaking into a car proves a pretty easy task, and many of your kind have done it successfully before, taking about twenty seconds to peel down a car door frame or claw open a window to secure a loaf of bread. Yet the cars in this lot are disappointingly clear—nary a Lifesaver or an M&M on the floor mats.

You're disheartened. The education has been too successful here as well. The Yosemite Lodge shows a video at check-in about your tendency to break into cars, and the reception staff brief all guests on the importance of removing food and other scented items from their automobiles. Room keys are not issued until guests sign an agreement on proper food-storage regulations.

6 a.m., Monday: Maybe those folks at Housekeeping Camp have gotten lazy in the middle of the night, you reason, and make one more circuit. You decide to head over to the Lower Pines campground in time for breakfast. On the way you cross the Merced River as the first light touches the top of Half Dome. At the other side of the river, you gaze at the oak trees, craving the acorn feast that will come in the fall, but for now the smell of scrambled eggs and bacon drifts in from the campground. At 6 a.m., you arrive at Lower Pines and do a circuit of its sixty sites, hoping to come upon the spoils of an absentminded camper. Again, nothing. You think you might give the place one final round, but then you notice a person wearing a flat tan hat and the unmistakable moss-green uniform of a park ranger. Your nemesis in your quest for human food, the bear patrol, has arrived. Someone must have spotted you and called the bear hotline. Time to retreat before the hazing begins.

A black bear portrait by remote camera trap.

7 a.m., Monday: This batch of visitors has been too well educated by park staff—you can't depend on them for food, so you
dash away from the people and head into the wilder areas of
Yosemite Valley, away from human occupation. You take a steep
climb to a level area above Staircase Falls and hang out for a time,
enjoying the quietude while munching on some shrubs. After the
meal, you nap contentedly, warmed into sleep by the late-morning
sun. In the early afternoon you race up to Glacier Point as deftly
as any rock climber, making the three-thousand-foot climb from
the valley below and taking in the magnificent view when you
stop to rest. You figure the tourists at Glacier Point have also been
too well trained to be irresponsible with their food, so you decide
to forage in a secluded forested area in between Glacier Point and
Sentinel Dome, then sleep though the heat of the day.

7 p.m., Monday: At dusk, you are ready to roll again, walking
around Sentinel Dome, then taking a short but steep route as you

head back down to Yosemite Valley, stopping briefly at 9 p.m. on the Four Mile Trail, devoid of hikers at this time of night. You follow your nose to the carcass of a newly dead fawn and relish the rich feast. After this meal, you stand upright and scratch your back up and down, up and down, on a nearby ponderosa pine, grunting with the exquisite satisfaction of curing an itch.

11 p.m., Monday: You cover the last five hundred feet to the valley floor, stopping at a wild onion patch for another snack, then amble past Southside Drive—lucky for you no cars in sight this late at night, as automobiles have hit more than two dozen of your fellow bears this year. You hang out near Swinging Bridge, where earlier in the year you ate many a grass salad. Today you dig up a nest of carpenter ants for a delectable insect feast.

5 a.m., Tuesday: You're getting full, but still, why not have another try at Housekeeping? Nothing. You also make another breakfast attempt at the campgrounds. Nothing there either. You proceed with caution in case the bear patrol is in the area.

7 a.m., Tuesday: You head over a creek and do a bit of foraging. Your appetite is satisfied for now—thankfully with all-natural food—so as the morning arises you decide to recover from your more-than-ten-mile journey and settle in for a long rest about a quarter mile from the Ahwahnee hotel. Although you didn't get to munch on any human food this time, you know on another trip you might be successful, as you have been before. You are hoping someone will be careless. As you drift in and out of sleep, secluded among the trees near the Merced River, the scent of fine cuisine from the award-winning Ahwahnee Dining Room wafts into your slumber, a temptation your ancestors certainly never had to contend with. You are part of a modern generation of Yosemite bears that dream of a cider-braised pork osso buco or Le Belge truffles prepared by world-renowned chefs.

> "The Black Bear will eat anything and everything it can lay its paws on."
> —Joseph Grinnell, *Animal Life in the Yosemite* (1924)

Yosemite Valley is like no other place in the world. An awe-inspiring landscape by all accounts, 150 years ago it provided the first impetus for the visionary idea of a protected national park, and today it attracts millions of visitors from around the globe. Called the "incomparable valley," the Yosemite landscape testifies to nature in peak artistic form; using her sculpting tools of glaciers and rivers, water and wind, she carved and polished spectacular rock formations (most famously Half Dome and El Capitan), and then decorated a handful of them with an array of waterfalls cascading over the high cliffs. As the late Steven Medley described in his popular guidebook to the park, "With its granite monoliths, towering waterfalls, and pastoral meadows, the valley is unique in the world for its remarkable scenery."

While representing just a small fraction of the twelve hundred square miles of Yosemite National Park, Yosemite Valley attracts the majority of the park's more than four million visitors each year. Within its seven-mile-long and one-mile-wide area it houses four campgrounds with more than four hundred sites, four hotels with more than one thousand rooms, and homes for an additional one thousand employees. For all its beauty, Yosemite Valley is in effect a miniature city. The hungry visitor can choose between twelve different restaurants—from the fine dining at the Ahwahnee hotel to the pizza shop at Curry Village—or go shopping at two well-stocked grocery stores. On a busy summer day, more than twenty thousand people may visit Yosemite Valley, sharing at least one thing in common—they all need to eat.

As does the black bear. And he needs to eat a lot, on average from four to six thousand calories a day, the equivalent of eleven Big Macs. During hyperphagia (the period in the fall preceding hibernation), bears go on a feeding frenzy in order to consume the twenty thousand daily calories they need to survive the winter; some bears even double their body weight during this time.

Black bears share similar tastes and nutritional needs with us humans—we both rank as omnivores, subsisting, at least ideally, on a balanced and diverse diet of mostly vegetable matter. That bears in fact do not live up to their reputation of being ravenous meat eaters surprises most people, and even more surprising is that most of their "meat" sources come from grubbing for insects. Yes, Stephen Colbert had it wrong in declaring bears the number-one threat to America and deeming them "godless, soulless killing machines." Most of the time, they stalk and hunt caterpillars or placidly munch on berries and flowers.

We also share similar digestive processes. Renowned bear expert Stephen Herrero observes in his book *Bear Attacks: Their Causes and Avoidance* that our cuisine attracts bears not simply because they will eat anything in sight but because our food and garbage easily convert into calories for them. Bears have never been picky about where their next meal comes from, and this adaptability has served them well over their evolutionary history, providing them with a variety of wild dishes to choose from in case of shortages. If a pine nut crop failed, they could seek out more insects. If the manzanita trees didn't produce berries in large quantities that year, wild cherries might make up the difference. So by tempting them with the smorgasbord of people food in Yosemite Valley, how could we expect this opportunist to resist three million years of evolution telling it to expand its menu options?

In retrospect, having let this three-hundred-pound kid loose in the proverbial candy store, we should not be surprised at the disastrous consequences that followed: property damage for us but often loss of life for the bears.

Steve Thompson and Kate McCurdy, the only two year-round staff members of the National Park Service's bear management team in Yosemite, were frustrated. It was 1998 and they were overworked, exhausted, and heartbroken. Both loved wildlife and bears. Both were dedicated to their jobs. Despite pulling double shifts—doing their administrative work by day and patrolling the campgrounds and parking lots as bear police at night—the situation continued to worsen considerably. "It was an impossible situation," remembers Kate. "I was just so tired of killing bears—it's not what I had gotten into this work for. I was willing to try anything to stop this."

Steve and Kate had a difficult job and, like their predecessors, did not have many options except euthanizing bears to stop what seemed like an unending problem. In 1998, despite extraordinary efforts from the entire park community over several decades, a record-breaking total of 1,584 human-bear incidents had occurred, the bears having broken into 1,354 automobiles and caused $659,569 in property damage, not to mention "wreaking havoc in the backcountry." As Kate told the *Los Angeles Times:* "There were bears everywhere. They were in the campgrounds getting into cars, into the trash; they were learning it from each other. My boss and I would drive all over the valley at night and you would just see the bears scatter."

But Yosemite Valley wasn't always a popular black bear dining hall. Although it is in the ideal range for these bears, who tend to prefer the lower- and mid-elevation forested areas in the park, usually below eight thousand feet, it doesn't offer an abundance of fare—it's basically a small garden protected by sheer granite cliffs. Before the influx of tourists and their moveable feasts, and before local eradication of the grizzly bear, who competed with the black bear for food and occasionally dined on the smaller bears themselves, Yosemite Valley probably supported about two to four of these animals. Today, around a dozen black bears frequent the valley, many just passing through, ursine carpetbaggers sustained by human food—a lifestyle that has altered the animals' behavior and put both people and bears at risk.

Feeding the bears, intentionally or unintentionally, has a long and infamous history in national parks, and Yosemite is no exception. In 1916, park rangers began staging shows for visitors, thrilling tourists by having bears eat from their hands. An official petting zoo opened in 1918, filled with an assortment of orphaned animals, including mountain lion kittens and one bear cub. Bears regularly begged by the side of the road, eagerly feasting upon handouts from visitors. During the 1920s and 1930s, a popular tourist attraction involved scheduled bear shows at the park's dump sites, where attendees willingly suffered the stench of burning garbage for a glimpse of a bear. Not surprisingly, around this time the bear population in Yosemite increased, and the first bear break-ins began in the park.

The situation quickly escalated in the many parks across the country—Sequoia and Kings Canyon, Yellowstone, and Great Smoky Mountains, to name a few—and during this time, park officials killed countless bears. To paraphrase a line often used about national parks, people were "loving the park bears to death," although for the most part the general public were unaware of the fatal consequences caused by their demand for entertaining ursine encounters.

The park shut down the petting zoo in 1932, discontinued the bear feeding shows by 1945, and closed many of the dumps, although some remained until 1971. Yet park visitation kept increasing and, along with it, the availability of human food. With their dump cafes closed, bears turned to securing meals from the more readily available food stored in campgrounds and cars.

The concept of bear proofing was nothing new. It had first been suggested and tried in the park in the 1930s, but as it turned out, the bears' intelligence rivals our own—once a bear figures out a new method for getting its snacks, it's back to the drawing board for the humans. Bear proofing is like trying to hit a constantly moving target; in that way, it is akin to needing to constantly update your computer's security systems to stay one step ahead of the viruses and hackers.

One inventive ranger imported lion poop from Africa, thinking the smell of a fierce predator might scare away the bears. Not only did this fail to intimidate the bears, but the UPS driver who had to handle the foul-smelling package threatened to cease all deliveries to Yosemite if they "tried that stunt again." Another researcher tested a method of electrifying cars in order to aversely condition bears to automobiles, but the bears simply, and quickly, figured out how to avoid the one car that delivered a shock. As proven again and again, black bears are very tough to outwit.

When visitors to the backcountry were urged to bear proof by hanging their food from trees, bears would chew through the ropes or climb the trees and bat at the food sacks like piñatas. Another technique, deemed "kamikaze bear," involved the animal leaping from a tree onto a food sack and then falling to the ground with the prize clutched in its paws. If the tree branches looked too flimsy to support the weight of an adult bear, the mother would send her cubs to take the jump.

Dumpster diving was another big problem. Although staff and trash collectors on morning patrols became used to the sight of bears sheepishly poking their heads out from underneath dumpster lids, the scene certainly startled many an unsuspecting visitor. One year, bears got themselves stuck in dumpsters more than fifty times and required assistance to get out. One memorable animal

A bear breaking into a car in Yosemite.

climbed on top of a dump truck and rode for more than a mile unbeknownst to the driver until a coworker alerted him to the presence of his hitchhiker. Other bears went undetected and were crushed in the dump trucks before they could be rescued.

Finally, J. R. Gheres, a foreman at the concession stables who was tired of the nightly racket the bears were making while raiding the dumpsters where he lived, created a carabiner retrofitted to lock the hatches. As Kate recounts the story in Rachel Mazur's *Speaking of Bears: The Bear Crisis and a Tale of Rewilding from Yosemite, Sequoia, and Other National Parks,* "This cowboy figured it out because he wanted to get more sleep."

As bear–human conflicts and property damage kept increasing, park managers, fearful for human safety and possessing limited options, continued to resort to lethal control. But unlike in the early history of the park, this time the problem became public and erupted into controversy. In 1973, the *Los Angeles Times* ran an article revealing that two hundred bears had been killed in just twelve years in Yosemite. The late photographer Galen Rowell also penned an article for the *Sierra Club Bulletin* titled "The Yosemite Solution to *Ursus americanus*" that fanned the flames. Public outcry was swift, prompting the adoption of a new management policy the next year that limited euthanasia, a move then followed in 1975 by the development of the first comprehensive bear management plan in Yosemite. The new measures were well intended but largely ineffective, as the park did not have the funds or resources to implement them.

In 1997, Steve Thompson sent a draft of the new bear management plan to ursine expert Stephen Herrero, from whom he received this alarming response:

> The number of bear incidents that occurred during 1997 place Yosemite in a class by itself in terms of bear problems....There is no doubt you understand the bear incident problem—bears getting into people's food and garbage. It is equally clear that you understand the solutions in terms of bear proof storage of food and garbage, and the human dimension of enforcing corrective action. It is time you be given the budget to do the job.

Getting that budget, however, would require an act of Congress, as national park funding is administered as part as the federal appropriations process. The trend had long been to shrink park budgets, not increase them, a situation that did not leave much hope for Steve and Kate.

That same year, faced with the heartbreaking task of euthanizing an entire bear family—a sow and two cubs had become aggressive because of park visitors yet again leaving out food—Steve decided to risk career suicide and picked up the phone. He didn't call another bear expert for advice. Instead he phoned the *Washington Post* and invited a reporter to come to Yosemite and witness firsthand the bears being put down. The story, "3 Bears

Too Clever to Live," told the tragic end of Miney and her cubs and considerably raised awareness about the plight of Yosemite's bears. The writer, William Booth, did not spare any details and gave the public an uncensored look at the consequences.

McCurdy squatted beside the bear and placed her hand on the animal's chest, feeling her heart. At first, it beat strong and measured, but as the poison moved through Miney's system, the rhythm quickened, became irregular, faster, then slowed, grew faint and finally could not be felt. And then the bear began to jerk and twitch. This took another few minutes. In a gesture old and unconscious, McCurdy held the bear's paw for a moment. She stood up, exhaled deeply and said, "Okay, let's do the cub."

The article caught the attention of the public, and a fortuitous visit to the park assisted even further. Phil Schiliro, who served as chief of staff to then congressman Henry Waxman, was vacationing in the park with his wife, Jody, a documentary filmmaker. They observed so much glass in the parking lots from bear break-ins, "you couldn't see the stars" because the shards were reflecting so much light. The couple attended a nightly bear patrol and checked cars for food right alongside the rangers. They even chased a bear from a garbage can.

Soon after the couple returned home, Steve received a phone call from Phil. "If we could help you, what would it take?" he asked. Steve and Kate answered, without much hope, that they needed $500,000, a number based on a recent program budget they had just developed. After a few months, Phil called again with the good news that an annual allocation of half a million dollars dedicated to solving bear problems in Yosemite had passed with little fanfare, and the park received the money in 1999.

Steve and Kate were giddy at the prospect of hiring additional bear technicians and adding much-needed resources. But even

more importantly, they knew this would help ensure a brighter—
and wilder—future for Yosemite's bears.

A black bear wandering Yosemite's Washburn Point, with Half Dome in the background.

The bears needed to be taught to be wild again. But it was the people who needed the education. And thus the National Park Service, with the help of all park partners (nonprofits, concessioners, and local businesses), launched the Yosemite Wild Bear Project, a comprehensive educational initiative aimed at the park visitor. In essence, it was an advertising campaign and, taking a cue from similar predecessors, like Smokey Bear's admonishment that "Only You Can Prevent Forest Fires," it appealed to people on an emotional level with the slogan "Keep Bears Wild." The campaign proved so successful it was even recognized outside the field— perhaps the first time a wildlife management initiative has been honored in *PR Week* magazine.

The praise was merited. Just one year after the appropriations, property damage from bears decreased by 54 percent, the cost dropping from more than $600,000 to $220,000.

The bear management team added multiple biologists, and the bear patrol increased its number of rangers at the height of the season. Bear canisters were now required for the backcountry, and an affordable rental program (five dollars a trip) helped make them accessible. The print shop kept busy producing "Don't Be Bear Careless" brochures, flyers, and posters that got distributed all around the park. Bookstores and retail gift shops across the park offered a complete line of products: stuffed toy bears, bear T-shirts, bear pins, and a poster proclaiming "A Wild Bear Is a Beautiful Sight to See." As visitors checked in at the hotels, the front desk showed video footage of bears breaking into a car. Employees for the National Park Service received intense training on tourist education. And the park also did outreach to the Yosemite gateway communities, asking for their cooperation and compliance as well.

The plight of the bears became impossible to ignore, as park visitor Steve Johnson related in an interview with the *San Jose Mercury News*: "You get hit in the side of the head with the education coming, going, and the whole time you are there."

The pressure didn't let up on the bears either. Bear proofing and deterrents continued, and became more effective with the efforts of the additional staff. New approaches were tried, such as the summer they recruited a team of Karelian Bear Dogs, whose fearless breed think nothing of chasing bears five times their size. Kate even acquired her own dog, Logan, and although he certainly displayed his ability to scare off bears, he proved his real worth as an ambassador: every camper who stopped to pet the adorable dog got an earful of bear education as well.

Other forms of hazing proved easier and cheaper than the dogs and have been a valuable part of the multipronged approach to conditioning the bears to permanently associate the developed areas of the park with "unpleasantness." In this effort, the patrols also fired rubber slugs from shotguns, let off noisemakers or "whistles, screamers, and bangers," and shined bright lights on the bears.

Visitors awakened in the middle of the night often complained, but the rangers used this as another teaching moment. Kate laughs when she remembers these episodes: "I'd often tell the disturbed campers, 'Go and gaze at stars—and try to marvel at how beautiful the park is this time of night. Then you can go back to sleep knowing you just heard the sound of a bear's life being lengthened.'"

The educational efforts, the deterrents, the bear proofing—all led to saving bears' lives. It's a myth that bears who eat human food won't go back to natural food; as researchers have observed again and again, even the most food-conditioned bears rely significantly on natural foods. The wanderings of the bear in the beginning of this chapter demonstrate this fact—he knew where the human food sources were from past experience, but when those sources didn't yield any meals, his wanderings took him to a nest of carpenter ants and a deer carcass. If people completely eliminated bears' access to human food, the bears would take Curry Village and the Lower Pines campground off their roving schedule.

"One of the coolest parts of my job is when we have successes with altering a bear's behavior from a negative to positive; adverse conditioning doesn't always work, but it can shift some bears," says Ryan Leahy, one of the bear biologists currently working in Yosemite. "One of my biggest successes was with a bear in Wawona breaking into eleven residences in the course of three days. He was going downhill and we did not hold up much hope." Ryan and his crew ended up tracking the bear and employed aversive conditioning on the animal for twenty-four

A bear cub in a tree in Yosemite.

hours, using noisemakers and shotgun rounds, making the usually peaceful campground sound like a warzone. The bear eventually climbed a tree, and Ryan and his crew got lawn chairs and sat under it, keeping up the unpleasant noises. Finally, the bear had had enough and started climbing down (at which point his human audience moved out of the way). When he was about fifteen feet from the ground, he jumped and darted into the woods, heading east.

A day later Ryan received a call from the forest rangers over in Mammoth Lakes, telling him they had observed a Yosemite-tagged bear "eating flowers in a meadow." Ryan laughs. "This is pretty crazy. This bear went from breaking into houses and eating human food to running forty miles over the Sierra Nevada and munching wildflowers in two days."

It's likely the bear will attempt to secure human food again. But he hasn't lost his wildness. He just needs people to remind him of his wild ways.

"No, don't despise the bear, either in his life or his death. He is a kingly fellow, every inch a king; a curious, monkish, music-loving, roving Robin Hood of his somber woods—a silent monk, who knows a great deal more than he tells."
—Joaquin Miller, *True Bear Stories* (1900)

"Selfies with Bears Prompt Warning from Park Rangers"
—2014 headline from *Digital Life*, National Public Radio

For me, every wild bear encounter has been like viewing an eclipse or a comet—seeing something elusive and beyond reach and that leaves you feeling like you've witnessed natural forces presenting you with a rare and precious gift.

The hike to Yosemite's Mammoth Peak, not to be confused with Mammoth Mountain, is wilder than it has a right to be, given that from Tioga Road you could almost toss a rock and hit its base. During one of my descents through a densely forested area crowded with willows and pines fighting over placement near the creek bed, I was lulled into the serene sense of being that is practically inevitable in this type of environment—enjoying the company of trees, the high mountain air, and the low music of the creek.

Suddenly a bear emerged from behind some trees and strolled by me in the distance. It seems we were equally surprised, startled out of our individual introspections. Perhaps the bear had been daydreaming as well, contemplating the richness of the berries he had just consumed, maybe savoring an aftertaste of caterpillars or anticipating the luxurious nap he would take later with a full belly. He was a gorgeous bear, his coat infused with sunshine—blondish, a light brown the color of fallen pine needles with cinnamon highlights. He turned his head to consider me, projecting the curious stare you might get from a person who was sizing you up.

A black bear in Yosemite.

Had he been human, he might have started asking questions. But
after a long moment, he gave a nod and resumed his ambling. I
watched transfixed as, like the final rays of light from the sunset,
he disappeared from view. Only upon reflection could I process
the surges of excitement and fear. I felt profound gratitude that I'd
witnessed this creature from such close range. There was a right-
ness to the encounter.

I could tell many Yosemite and Sierra bear stories from my
thirty years of wanderings. I've seen a bear stretch his mighty arms
and legs to climb a cliff below Muir Pass, his muscles rippling
beneath his fur coat, his paws gripping the rock and propelling
him up the nearly 90-degree slope. On the Chilnualna Falls Trail
one spring day, I watched a bear scratch his back on a tree, grunting with satisfaction, and after he had left, I made my way to the
tree and touched some of his fur left in the bark.

My favorite Yosemite bear story doesn't belong to me but is
one I stumbled upon in my research. If you look on a map you'll
see a notation for Lost Bear Meadow on Glacier Point Road, a

little east of the Badger Pass ski area near Bridalveil Creek. The tale goes like this: In 1957, a little girl, Shirley Ann Miller from Reseda, California, became lost while camping with her parents at Bridalveil Creek and possibly entertained a visit or two from a bear. She remained lost for three days and three nights, until rangers found her about two miles away sitting, unconcerned, on the side of a hill, gripping a spoon that she used to dip in water to drink. The ranger reported that when they approached her, she replied, "I am not lost, but the bear is lost."

And in this tale I am telling, I don't want the bear to get lost either—the true character of the wild, beautiful bear.

In wading through the news reports, historical documents, plans, and scientific papers to write this chapter, it struck me that it's very easy for the bear himself to get lost in the commotion of modern life; I don't want to mischaracterize him or leave those unfamiliar with his kind with the sole impression he's a dumpster-diving, campground-marauding thief whose only mission is to come for your cooler filled with lunch meats. For inspiration, I called up Kate McCurdy, who has moved on from Yosemite National Park and now directs the Sedgwick Reserve for the University of California. We remained friends after my almost decade-long stint working in Yosemite, where I served as one of the original members of the Yosemite Wild Bear Project. While we talked, I shared my feeling of drowning in bureaucratic readings and losing sight of the real wild *Ursus americanus.* "Forget the science and planning," I said. "What is it you really like about bears?"

She paused for a moment, then answered, "Of all the animals I have worked with, they appreciate beauty. You see them on their backs looking up at the blue sky and rolling around in flowers. They like free time. They like to lounge around in a mud hole. Maybe other animals that have to work harder to survive can't enjoy this freedom, but for bears, being on top of the food chain, like us, sometimes gives them the luxury of just enjoying life."

We share a special affinity with bears, perhaps because of these humanlike antics and personality traits, and a bear sighting, whether in a Los Angeles suburb or Yosemite Valley, inspires a whole slew of emotions, from fear to awe. For the most part he is irresistible to us. One bear made global news when he awakened from hibernation early and dashed across the ski runs at Heavenly Ski Resort during the 2014 World Cup competition. When the bear named Meatball roved the neighborhoods of Glendale (see page 33), his adventures were tracked by a helicopter film crew. Two bears cubs playing on a road in Yosemite caused a traffic jam that lasted for hours, and the video raced across social media. The public dearly loves a bear tale.

The challenge is to find the balance between feeling a connection with them without interfering in their wild lives—or putting our own in danger. Perhaps the best example of our affection for bears run amok is a story that seems straight out of the satirical news source *The Onion*. In 2014, a startling social media phenomenon surfaced: the #bearselfies fad had grown so much that officials were issuing warnings and pleas for people to stop taking self-portait photos with bears. As journalist Benjamin Spillman reported in the *Reno Gazette-Journal,* "We've officially arrived at the point where people need to be told taking 'self-ies' with bears is a bad idea." The trend originated in South Lake Tahoe, where the annual run of kokanee salmon attracts hungry black bears. All it took were a few visitors snapping photos of themselves with bears in the background and the craze spread across Facebook, Instagram, and Twitter. Some even made video #bearselfies. Though the underlying motivation for the trend is heartening—it's an acknowledgment of the specialness of these encounters—encouraging a bear to photobomb you is terribly irresponsible. You're putting both yourself and the bear at risk. For those considering seeking out a selfie opportunity with bears, you might want to consider Theodore Roosevelt's observation: "My

A black bear feasting on a carcass.

own experience with bears tends to make me lay special emphasis upon their variation in temper."

In California, black bears are more commonly becoming a part of our lives—even in urban and suburban areas. As our population increases along with the bears', our interaction doesn't have to be a collision course of shattered windows, scattered garbage, and, ultimately, euthanized bears. As with mountain lions in the heart of Los Angeles, peaceful coexistence is possible. The success of "Keep Bears Wild" is hard won but encouraging, and this philosophy is exportable beyond the park's boundaries. The California Department of Fish and Wildlife has also developed a Keep Me Wild campaign and declared May "Be Bear Aware Month," prompted by increasing incidents of bear-human interactions. The CDFW receives calls whenever bears break into homes, rummage through trash bins, and raid campsites. "These bears are often labeled 'nuisance' bears," says a CDFW press release, "but in reality they are just doing what comes naturally to them." We need to

Keep Bears Wild throughout all of California if we want them to have a future in our state.

For all this talk of marauding bears in garbage cans and of the perils of us habituating bears to our lifestyle, I also want to caution us against habituating ourselves to a bear of our own creation. Taming the bear to meet our needs is akin to taking a chainsaw to a giant sequoia or a wrecking ball to Half Dome. We don't want a whole generation of children—and adults—to think it's normal for bears to pop open our trash bins or break into our cars. There is something not just unnatural but undignified and tragic about seeing a bear munching on the remains of a bag of chips. I feel ashamed of us for robbing him of his wildness.

A wild bear is a beautiful sight to see. My sincere wish is that everyone has the opportunity to witness a wild bear doing wild bear things: munching on a dandelion, scratching its back on the rough bark of tree, chasing a hapless butterfly, or rolling in the mud.

Leave the bear selfies to the bears.

In March of 2014, Yosemite's bears once again made the news, but this time the stories signaled cause for celebration. A new study had just been released showing the bears in Yosemite now consumed natural foods at levels more appropriate to a wild ursine, evidence that they had curbed their late-night snacking on potato chips and soda. Comparing historical samples of bear furs and bones to recent samples, researcher Jack Hopkins, the lead author of the study and a former member of the Yosemite bear team, found that the proportion of human food in the bears' diet had shrunk to 13 percent, a level not seen in the park since the early 1900s. As he told the *Washington Post,* "What we found was that the diets of bears changed dramatically after 1999, when the park [developed a plan] to keep human food off the landscape."

All of the efforts over the years had not been in vain. Aside from the bears' return to wild foraging, the statistics also demonstrated just how much progress had been made. Since the launch of the Keep Bears Wild education campaign, bear-related incidents have decreased by 95 percent, and in 2015 the park recorded the lowest number of human-bear interactions since record keeping began in 1975.

But despite the progress, there are always new visitors to educate, and so the job is never done. Caitlin Lee-Roney, one of the bear biologists currently in Yosemite, knows that you can never hang up the hat. "People think Yosemite is doing so great that we have fixed our problem," she says. "But [that's] almost when you need to push harder. As soon as you stop, the bears will resume their unnatural habits. As long as human food is in the valley, they are always going to try and obtain it. My passion for bears and not

Yosemite bear biologist Caitlin Lee-Roney.

wanting a repeat of the past is what keeps me going, even when you have to tell people two hundred times a night, every night, why leaving a cooler on a picnic table unattended is bad." As biologist Ryan Leahy notes, "It only takes one incident, one unattended cooler in a car, to lose important ground with a bear."

Tom Medema, Yosemite's chief of interpretation, tallies up the enormous effort his department has to continually undertake to keep visitors educated. "Each year interpretive rangers dedicate over one thousand hours focused specifically on educating campers and picnickers about food storage," he says.

Another tool has been added to the repertoire—a new technology that is beginning to help immensely with this work. In 2014, the Yosemite Conservancy donated $70,000 to purchase fifteen GPS trackers for the bear program. (In the last twenty-five years the group has donated more than $2.1 million to help with bear issues in the park.) Tracking bears in real time and across distances not covered by radio telemetry has enabled the bear team to more efficiently divert resources, meaning they can deploy staff when and where bears are active rather than just responding after the fact.

Aside from increasing use of the latest technology, the park has also expanded the Keep Bears Wild program in other ways, and widened the overall educational message. Rachel Mazur, Yosemite's chief of wildlife, points out, "It's not just the bears that get into our unsecured food and trash and shift their behaviors to focus on it. Ravens, squirrels, raccoons, and coyotes all do the same thing, and all have increased conflict with humans because of it. And it isn't just the big food rewards that matter. Micro-trash can be just as bad. Ravens, for example, are adept at finding small bits of food trash. This has led to an increase in the population, plus a range expansion, of these corvids, which in turn leads to a decrease in the less-dominant native birds that they edge out. If everyone spent fifteen minutes scouring their campsites for bits of trash before they left, it would make a world of difference."

Although it may seem counterintuitive to the overall goal of keeping wild animals wild, all of these human efforts are necessary in order to preserve the integrity of wildlife, whether in an urban setting or in Yosemite. Their fate now depends on us. As the *Washington Post* reported in 1997, "Though many visitors do not realize it, the wild places and wild creatures Americans treasure have become increasingly managed by man—down to the smallest detail."

Yosemite is one of the best-protected places on the planet, but even there the wildlife needs a little help from its human friends. Yosemite teaches us that wherever we live, keeping wildlife wild is our responsibility. What if all four million visitors arrived to Yosemite already trained in the tenets of Keep Bears Wild? What if we were all as aware of our garbage disposal in our backyards as we were in the park? The consequences are no less dire. Countless bears and coyotes, lured by pet food and trash, are killed across the country daily, along with raccoons, possums, skunks, and an entire menagerie of critters that share our space. By making our food accessible, we invite these animals into our neighborhoods, and many suffer fatal consequences. Yet the fixes are so simple. We're all part of the problem, but we're also all part of the solution.

The Yosemite Miwok tell a tale of two wayward bear cubs that ignore their mother's advice and wander too far from her side, splashing and playing and swimming in the river to their heart's content. They grow tired and nap on a nearby boulder, but while they sleep, the rock rises, "until the little bears scrape their faces against the moon," and they wake up to find themselves trapped, unable to descend down the dizzying heights to return home. All of the creatures of the forest miss the two bear cubs and make plans to try to rescue them, yet all fail, even the mighty mountain lion. Finally, a tiny inchworm volunteers, and despite the other animals' taunts, he starts to creep up the rock. His pace is slow, but with patience and courage he rescues the cubs by leading them back to the valley floor. The rock was then named in his honor, Tu-tok-a-nu-la.

Next time you visit Yosemite, take a look at his rock, now called El Capitan, and think of the tiny animal's long journey to save the two bear cubs. Surely we can match the efforts of an inchworm.

Note: My special thanks to Rachel Mazur, whose excellent book Speaking of Bears: The Bear Crisis and a Tale of Rewildling from Yosemite, Sequoia, and Other National Parks *provides a wonderful, in-depth oral history of the Yosemite bear management program and provided historical information and several quotes used in this chapter.*

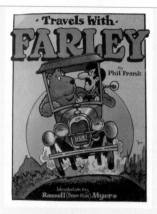

Phil Frank:
A Cartoonist Advocates
for Yosemite's Bears

Yosemite's bears had a champion in the late Phil Frank, cartoonist for the *San Francisco Chronicle*, as his daily comic strip raised awareness about the park's Keep Bears Wild program.

Featured in the *Chronicle* from 1986 until Frank's death in 2007, *Farley* included a storyline about the exploits of four ursine characters—Bruin Hilda, Franklin, Alphonse, and Floyd—residents of San Francisco who summered in Yosemite. Phil would often work with the bear research team for input on his content, and he even accompanied them on patrol in 1998. He genuinely cared for Yosemite's bears, and his strips' educational message—combined with a humorous approach—focused on the importance of proper food storage.

If you are visiting Cavallo Point in Sausalito, be sure to stop by the Farley Bar, named for Phil and decorated with many of his cartoons. Raise a toast to his photo hanging over the fireplace, and thank him for campaigning for Yosemite's bears.

This page: The *Farley* comic, featuring Yosemite bear characters.
Opposite page: A grizzly bear on the California landscape.

The Return of the California Grizzly?

California once offered "the richest, most diversified grizzly bear habitats" in the country, according to bear expert Doug Peacock. Reports from settlers and explorers in the 1800s described seeing the bears in "herds" and frequently witnessing "fifty or sixty within twenty-four hours." By 1924, however, the ubiquitous grizzly had vanished from the Golden State, destroyed by hatred driven by misperceptions about predators. Today, the grizzly bear occupies just 4 percent of its historic habitat in the United States, mostly in remote places such as Yellowstone and Alaska.

The prevailing modern philosophy of how to manage grizzlies involved simply maintaining the populations that remained. Aldo Leopold challenged that view in his 1949 book, *A Sand County Almanac*: "There seems to be a tacit assumption that if grizzlies survive in Canada and Alaska, that is good enough. It is not good enough for me....Relegating grizzlies to Alaska is about like relegating happiness to heaven; one may never get there." Today, Noah Greenwald, of the Center for Biological Diversity, shares Leopold's sentiment. His group submitted a petition to the US Fish and Wildlife Service in 2014 to return grizzly bears to larger portions of their historical range.

His motivation isn't just about restoring a lost heritage to California and other places but about ensuring the ultimate survival of the bear. The roughly eighteen hundred grizzly bears now living in the contiguous United States needs to be increased, as Noah wrote in the *LA Times*. "The small number of grizzly bears is simply not enough to ensure the long-term survival of the grizzly in the face of climate change and the ever-growing footprint of human development," he says. It's a challenging proposition, especially since traditionally prime grizzly habitats are now also some of the most densely populated areas in California. Not everyone agrees with his ambitious campaign, however. Although the idea of restoring a native species—and one as magnificent as the grizzly bear—in California is wonderful to contemplate, other scientists think that given the modification of habitat and increases in human population, it's not a realistic scenario. Noah remains undaunted: "It won't be easy, but all the facts suggest it can be done."

Slow Down for Yosemite's Bears

Although Yosemite National Park has enjoyed widespread success with enlisting people to Keep Bears Wild through proper food-storage techniques, efforts to eliminate another human source of bear mortality—speeding—has proven much more challenging. In 2014, twenty-five bears were hit by cars in the park, while in 2015 the number increased to thirty-four. The park has installed warning signs in key bear-crossing areas and augmented its educational efforts on the issue as well.

"Roads are a huge source of mortality for wildlife in Yosemite, especially when roads intersect wildlife travel corridors," says Rachel Mazur, Yosemite's chief of wildlife. "But there are solutions. We are installing large culverts for animals to cross under roads, putting in warning signs where they commonly cross, and reducing speed limits where visibility is limited. Now we just need visitors to slow down and stay vigilant for animals on the road."

Top: A sign in Yosemite warns people to slow down for bears.
Bottom: A bear crossing a road in Yosemite.

Bighorn Sheep Get a Helping Hand

Superintendent Don Neubacher possesses a lifelong connection to Yosemite, and he has his own bear story. Recalling the time a bear approached his family during an annual childhood camping trip, he says with a laugh, "My mom scooped me up and made a run for it." One Yosemite wildlife sighting he did not collect as a child, however, was that of a Sierra bighorn sheep; the animals disappeared from the park in 1914 and were not seen again in Yosemite until 1986, after they were reintroduced into Lee Vining Canyon.

During his tenure, Don has championed a project to increase the bighorn population of Yosemite. The National Park Service, with funding from the Yosemite Conservancy, released a new herd into the park in 2015. "This is such an inspiring project—restoring bighorn sheep to Yosemite's Cathedral Range after an almost one-hundred-year absence. I'm fortunate to be serving as the superintendent during such a momentous time," he says.

The Cathedral Range, famed for its spire-like peaks, provides plenty of ideal territory for the bighorn, such as cliffs that offer protection from predators and plenty of forage. In fact, this endemic subspecies evolved in the granite of the Sierra Nevada, and as longtime bighorn researcher John Wehausen once wrote, "Were granite to come to life, it would undoubtedly look like a bighorn sheep, so perfectly do they blend into their habitat." Located in the interior of the park, this area also reduces the risk of exposure to domestic sheep and the diseases they carry.

Biologist Sarah Stock, who has been working on the project team for five years, is proud of the effort: "It is remarkable to have them back in the heart of the park's wilderness. This restoration effort reminds us that humans are capable of reversing mistakes from the past. When you see the sheep up there, it will feel good."

Bighorn sheep being released in Yosemite in 2015.

How Does a Fisher Cross the Road?

In 2011, fisher expert Rick Sweitzer posted an unusual request on social media: Send me your spare socks. Tired of spending scarce research dollars on the two thousand socks he needed each year to enclose bait for his camera traps, he appealed to the public for donations.

The response was overwhelming. Media outlets across the state picked up the story, and one ran the headline "This guy wants to eat your socks" with a photo of an endearing adult fisher. Anne Lombardo, a coordinator on the project, remembers the hundreds of packages that arrived with more socks than they could use in a lifetime. "I thought we'd get a dozen," she says. "It's great to know that people across the state cared about this seldom-seen creature that they had probably never heard of before this."

Their work is part of the larger Sierra Nevada Adaptive Management Project, for which fishers are a species of concern. Fishers have disappeared from 50 percent of their range in California. When the study found that vehicles along Highway 41 were a significant source of mortality for the fisher, researchers on the project knew they had to find a solution. This time it wasn't as easy as getting people to donate spare socks, but it still came down to small fixes that can make a big impact. After camera traps showed that fishers were already using preexisting culverts to cross the busy road, a cooperative effort by Yosemite National Park, the US Forest Service, Defenders of Wildlife, and Caltrans is now identifying simple solutions to keep fishers safe, such as installing platforms in the culverts that rise above the flowing water—a modification that helps a variety of animals use these underground crosswalks year-round, meaning that automobiles may be one threat the fisher and many of his friends can now safely avoid.

Craig Thompson, another researcher working on fisher projects in the area, applauds his colleagues' efforts. "The fishers are facing so many challenges—loss of habitat, vehicle mortality, and even widespread exposure to lethal poisons from illegal marijuana camps," he says. "The least we can do is help them cross the road." Craig is working on other conservation measures, including a release program to return fishers to some of their traditional range and the cleanup of toxic chemicals found at illegal marijuana grows on public lands. "Given the volume of new threats facing wildlife in the Sierra Nevada region these days, cleaning out and retrofitting old culverts is an amazingly simple, cheap, and effective conservation effort," he says.

A Plea for the Pika

One of my favorite places in Yosemite, known as Tioga Pass country, offers stunning views of imposing granite peaks. Yet my affection for the area has less to do with the spectacular scenery and more to do with a small, furry, and almost ridiculously adorable creature less than eight inches long who could easily be overlooked amidst the enormity of the landscape.

The American pika, a small relative of hares and rabbits, lives in rocky terrain typically at elevations of eight thousand to thirteen thousand feet in California. Also fondly referred to as rock rabbits, boulder bunnies, or "little chief hares," pikas are often heard rather than seen and they call with a distinctive high-pitched "meep!"; appropriately, the animal's name may be a derivative of the Russian word *pikat,* meaning "to squeak." Despite the harsh winter conditions they experience, pikas do not hibernate but instead collect various grasses, shrubs, and lichens, place the food in the sun to dry, and then stash it into a "haystack" for winter consumption. These nimble and deft creatures have also been known to loot their neighbors' haystacks.

Climate change is having an impact on the cold-loving pika, although not uniformly across its range—some populations remain steady while others have vanished. Scientists including the US Geological Survey's Erik Beever have tirelessly researched many aspects of pikas to best inform the species' management and long-term conservation range-wide. As a volunteer helping in the research efforts to study pikas, I've spent years with Yosemite's pikas, snapping photos and submitting data on sightings. And I am just one of hundreds of people volunteering their spare time to help the pika across the country; numerous other citizen science groups have formed to help monitor this imperiled critter. "In the arenas where we're working our research is enhanced greatly by the citizen science efforts," says Erik. With the dedication of such an impressive network, there might be hope for this diminutive creature yet.

This page: A pika in Yosemite's high country.
Opposite page: A fisher kit in Yosemite.

Great Gray Owls Go Digital

"I am obsessed with raptors in general, and for me this is the ultimate raptor," says Joe Medley, a UC Davis PhD candidate who studied the great gray owl for years in Yosemite and other areas. "They are the largest of the North American owls. They manage to survive winters in Yosemite. They can hear a mouse under a foot of snow."

Early biologists who visited Yosemite also noticed the uniqueness of the bird, including Joseph Grinnell and Tracy Storer, who in their 1924 book, *Animal Life in the Yosemite,* wrote: "The discovery of the Great Gray Owl in the Yosemite section was one of the notable events in our field experience. And what was most surprising was the fact that the bird was apparently quite at home, and nesting. No previous record of the breeding of this northern species of owl south of Canada is known to us."

One unique method that Medley developed when he worked in Yosemite helped speed up data analysis. Using digital audio recorders and voice recognition software to track the birds, Medley designed an algorithm to more rapidly process data from the recordings, speeding up a process that would otherwise have taken seven years of listening. "It's pretty cool that we can use this technology as a supplement to traditional monitoring techniques," he says. "It's really cost effective."

Yosemite's great gray owls are rare, and the species is listed as endangered in California. The park provides a last haven for these birds and is home to the majority of the entire state's population. Research conducted by Medley and others, including park biologist Sarah Stock, is vital to understanding the long-term health of a population that has been evolutionarily distinct since the late Pleistocene. With increasing loss to the bird's habitat from wildfires and development, this research has become more vital than ever.

This page: A great gray owl in the Sierra Nevada.
Opposite page: Yosemite toads in amplexus.

The Love Song of the Yosemite Toad

Let us celebrate the Yosemite toad—for his sonorous musical trilling that matches any birdsong in spring, for being an intrepid amphibian who survives in the alpine meadows of the Sierra Nevada, and, as George Orwell observed in his eulogy for spring, "because the toad, unlike the skylark and the primrose, has never had much of a boost from poets."

Californians should take pride in the Yosemite toad; it's a native son found nowhere else on earth except the high elevations of the Sierra. Mountain life isn't an easy existence for amphibians, and the toad spends half the year in hibernation, but once the snow melts—or even before, as the critter has been observed tiptoeing over snowfields to reach its breeding grounds—the males emerge from their sleep and find a suitable pool to begin their annual search for mates. The toad's distinctive love song can be heard up to one hundred yards away, and as naturalists Grinnell and Storer noted in 1924, "Its mellow notes are pleasing additions to the chorus of bird songs just after the snow leaves."

Once present in abundance, the amphibian pride of the Sierra is now disappearing from its home. Yosemite toad populations have vanished from areas of their historical range, and researchers and government agencies are taking steps to protect this threatened animal. Rob Grasso, a biologist with the National Park Service, advocates for the protection of this special creature. "The Yosemite toad is the most captivating species I have ever studied," he said. "I can tell you everything we know about them, and then tell you that pales in comparison to everything we still don't know about them. They are truly fascinating animals that keep their secretive ways well hidden, which to me makes them all the more extraordinary [and important] to preserve." In 2014 the US Fish and Wildlife Service granted threatened status to the animal, and researchers in Yosemite, like Grasso, continue to document critical habitat and attempt to determine causes for the population's decline so they can reverse this trend and ensure that the love song of the Yosemite toad is heard long into the future.

A gray fox named Squat, in Silicon Valley.

CHAPTER 4

Friending Wildlife

THE FACEBOOK FOXES AND WILDLIFE CORRIDORS IN SILICON VALLEY'S HIGH-TECH WORLD

"The Facebook foxes that live on our campus are pretty amazing. It makes me happy that we got our campus certified as an official wildlife habitat so these guys could stick around."—Mark Zuckerberg, Facebook CEO

"A fox never found a messenger better than himself."—Irish proverb

In June of 2013, a gray fox pup emerged from nearby bushes and peered into a window. The business discussions taking place in Mark Zuckerberg's office at Facebook's Menlo Park campus suddenly ceased. As the young fox with its curious eyes, pointed ears, and black-spotted muzzle considered him through the glass, the social media pioneer whipped out his smartphone and began snapping photos. Andrew "Boz" Bosworth, Facebook's vice president of advertising, did the same, capturing Mark in his characteristic plain T-shirt beaming a smile at one of the adorable new campus mascots. "FB Fox crashed Zuckerberg's meeting" read the caption when Boz shared the picture online.

From there, the news spread. The blog Gawker announced "Mark Zuckerberg 'Likes' Something That's Awesome: Baby Foxes." The *Huffington Post* declared "And they're the cutest. Little. Creatures. In the World." Even the *Wall Street Journal* took a break from pontificating about the one-year anniversary of Facebook's IPO and reported "Baby Foxes Check Into Facebook."

The pup was the offspring of a female fox that had first been spotted wandering in Hacker Square about a year before. Glimpses and rumors gave way to confirmed sightings, and soon three new pups rounded out a "skulk"—the proper collective noun—of Facebook foxes. Employee and enthusiast Alexis Smith set up a page, "FB Fox," about a month before the famously crashed meeting. "Everybody was posting photos to their personal pages—I wanted to create a space we could honor the foxes together," she said. After Mark liked the page, it gained three thousand fans in twelve hours. After the media picked up the story, page subscriptions skyrocketed. Today it boasts more than one hundred thousand followers from forty-four countries around the world.

As fully as any sports mascot at a university, the foxes became embedded in the DNA of the campus. The employee store offered stuffed toy foxes sporting Facebook monogrammed shirts, the staff proudly wore "Fox Club" buttons, and the company commissioned an artist to paint a fox mural in one of the buildings. Alexis even championed the creation of Facebook fox emoticons—a series of foxes brandishing the thumbs-up sign, a pint of beer, or a happy face—for all your metacommunicative needs.

People could not get enough photos of the foxes dashing across the basketball court, relaxing on the picnic benches, playing tag on the lawn, or simply trotting through an outside work area, casually passing employees coding away on their laptops. Technology humor abounded on the FB fox page, such as when staff person Steve Kaye

Fox pups playing on the Facebook campus.

Foxes pose at the famous 1 Hacker Way sign.

shared a photo of the mother fox, nicknamed Firefox after a popular Internet browser, strolling on the campus carrying a meal of a dead bird in her mouth; the caption read "Mama F putting an end to all the tweeting," a dig at the social media platform Twitter. The page also featured an entire series of photos of foxes napping contentedly on the hoods or roofs of the employees' cars, with captions like "BMW: The ultimate driving and napping machine."

Perhaps the most endearing shots depict what makes the animal distinct from its other fox relatives. Tamara Eder, author of *Mammals of California,* describes them thusly: "Truly a crafty fox, the Common Gray Fox is known to elude predators by taking the most unexpected of turns—running up a tree." The FB foxes darted up trees and climbed the awnings above campus walkways as people passed underneath.

To ensure the foxes' health and safety, Facebook's facilities management team worked with wildlife services and contacted local gray fox researcher Bill Leikam, who gave a presentation to the employees designed to educate them about the animal. Facebook employees initiated additional steps to guarantee that these wild creatures stayed wild, an encouraging echo of Yosemite's efforts to educate the public about healthy interaction with black bears. The motto of the FB fox page became "Please honor the foxes—no chasing or feeding—just mutual respect." Staff posted signs on campus asking people to keep their distance. An employee shared a photo of one of the foxes resting under his automobile and warned, "Before you drive off, please check if there is a fox under or near your car."

Bill, known by many as the Fox Guy, accompanied me on my tour of Facebook's campus to witness the foxes for myself. After we posed for the requisite selfie at the famous 1 Hacker Way "Like" sign, we met Jacqueline Rooney from corporate communications in the front lobby for our tour. Once inside, I no longer had any questions about why a fox family would want to take up residence here—I now wanted to live on the Facebook campus. Designed by consultants from Disney, the campus exudes the "happiest place on earth" vibe. Tree-lined pedestrian and bike paths pass the Sweet Stop ice cream parlor, a pub, and an assortment of cafes, most free to the employees. The array of amenities also includes a bank, a bike shop, a barbershop, a health care center, a music studio, and a woodshop. The campus is saturated with greenery, from a community garden to abundant plots of trees, flowers, and native grasses. Employees program on comfortable lounge chairs and outdoor sofa sets, periodically rising to do tai chi stretches and yoga poses on the lawn.

Alexis met us at the Zen Garden, an expansive deck in between two buildings surrounded by a mulched garden area of trees and flowers. She pointed to the space underneath the deck where the foxes denned and gave birth. She recalled when she first viewed the pups romping here, wrestling and chasing each other's tails; their antics made it easy for employees to become smitten with the three "foxeteers." Alexis also recounted that the pups "would nurse right here on the boardwalk as employees walked by."

Unfortunately, the foxes were nowhere in sight that day, so we chatted about our favorite fox antics (my vote was for tree climbing), and Bill answered some questions about fox biology (they are able to climb because their forearms rotate). Before leaving, I presented fox-friendly Facebook with the National Wildlife Federation's official Certified Wildlife Habitat designation sign (see page 178), which was proudly received. The next day, Alexis posted to the "FB Fox" page a photo of the sign, already mounted on a fence post in the community garden, a favorite hangout of the foxes. About a week later, above a photo of two fox pups curled together while napping, Mark commented to his tens of millions of followers the epigraph to this chapter.

In that one post, Mark did more to promote awareness about urban wildlife coexistence than I could ever hope to achieve in my career.

Aside from fueling a social media #cuteoverload #fbfox trending topic, I began to think about what all this meant—for the foxes and all urban wildlife. One of Mark's goals is to "connect everybody." Could this "everybody" apply also to wildlife? The Facebook staff—including those more schooled in HTML and JavaScript than in the principles of conservation—had grasped the whole coexistence thing with their usual savvy and quickness. The staff rapidly integrated "wildness" into their campus, and they did it without endless committee meetings and hundred-page employee manuals of rules and regulations, and nary a task force to guide them. In its championship of the skulk of foxes, Facebook had pioneered yet another frontier in Silicon Valley, marking a milestone for human–wildlife coexistence. Had social media, and the culture that created it, changed the game? Did we ultimately owe this embracing of foxes to crowdsourcing?

In my quest for forging new paths for urban wildlife conservation, I started concocting possibilities. BuzzFeed had asked in one of its features on the Facebook foxes, "Ugh, why can't all offices have a family of foxes living in them?" My mind began racing. Should my new conservation tactic be releasing charming wildlife onto high-tech campuses? Maybe some bobcats to tempt Google's CEO? Or beavers for Apple's leadership?

Mark Zuckerberg ✔ shared Josh Frankel's photo.
June 28, 2013 ·

The Facebook foxes that live on our campus are pretty amazing. It makes me happy that we got our campus certified as an official wildlife habitat so these guys could stick around. Check out the FB Fox page for more adorable photos.

👍❤️ 166K   155 Comments   8.9K Shares

👍 Like   💬 Comment   ➡️ Share

"My conception of grandeur, beauty, and commercial magnificence is realized
in the Santa Clara Valley."
—Edward Jeffery, president of Western Pacific, 1913–1917

"And yet, as the humans eat dosas and climb fake mountains and learn
acupuncture and buy lap dances, beneath the asphalt and concrete, the
microbes eat toxic waste sweetened with molasses, cleaning up our mistakes."
—Alexis Madrigal, *Atlantic Monthly*

That foxes romp anywhere in today's Silicon Valley
seems a remarkable feat given the environmental history of the
region. Like the porpoises in San Francisco Bay, it required restoration—intentional or unintentional—before wild things would
venture back into some areas. Among the early settlers of Santa
Clara Valley who recorded its incredible beauty was William
Brewer, who in 1861 wrote, "The Santa Clara Valley is the most
fertile and lovely of California." Once word of its allure spread,
the inevitable followed; as more recent chroniclers of California
history, the Eagles, warned in a song, "Call someplace paradise, kiss
it goodbye."

First the cattle ranches arrived, their animals sustained with
plentiful grazing on hill and dale. Next, the arrival of the railroad
and the discovery of abundant artesian well water attracted agricultural interests, and the livestock were soon replaced by extensive orchards of apricots, cherries, and plums. Each spring, tourist
publications touted "blossom tours" to lure visitors to "the Valley
of the Heart's Delight," the region's new nickname. Population
and development subsumed the native landscapes of oak-lined
hillsides, fresh- and saltwater marshes, expansive grasslands, and
naturally flowing rivers and creeks. And with this incremental
banishment of natural space came a gradual loss of wildlife. Over

time the plentiful herds of tule elk and pronghorn vanished, followed by the grizzly bears, beavers, and otters. The once numerous flocks of shorebirds and waterfowl significantly diminished.

Yet the environmental impacts of cattle ranching and fruit farming paled in comparison to the rapid degradation that befell the region during Silicon Valley's industrial age, which started in earnest after World War II. The tinkerings of Stanford graduates William Hewlett and David Packard in their garage (which is now a private museum) heralded Silicon Valley's birth as the center of the high-tech universe, but unlike the largely service-oriented economic base of today, manufacturing then reigned as the dominant industry. Companies including National Semiconductor, Xerox, IBM, Atari, Apple, Intel, and Hewlett-Packard actually built their products within Silicon Valley's borders, instead of building them overseas as they do today.

Silicon Valley's orientation toward technology manufacturing required paving over much of the land to meet the rampant demands for commercial space and housing; more than two hundred thousand jobs were added during the mid-1960s to mid-1980s, and those people needed places to work and live. Meanwhile, toxic chemicals associated with the manufacturing process contaminated the land and water, rendering many places uninhabitable for flora or fauna. Today, Silicon Valley ranks as one of the most toxic areas in the country, with almost two-dozen Superfund sites. With the double whammy of poison plumes and land sealed off with asphalt and concrete lacquering, the local wildlife's future didn't looking promising.

A group of mothers concerned about chemicals leaking into the groundwater raised an alarm in 1982, and in response two attorneys formed the Silicon Valley Toxic Coalition to strengthen regulations and advocate for mitigation efforts that are still underway to this day. Although Silicon Valley is now largely devoid of manufacturing plants, the toxic leftovers still pollute thousands of

One of the Facebook foxes on campus.

gallons of water, which must flow through extensive treatment systems to exorcise the chemical ghosts.

Above ground, the grassroots open-space movement raised funds to purchase land, lobbied for changes in zoning laws, promoted conservation easements, and formed a new type of agency for land management called Open Space Districts. And it worked. Daniel Press, the author of *Saving Open Space: The Politics of Local Preservation in California,* acknowledges the effectiveness of this movement: "For all the ways in which the Golden State resembles and differs from the rest of the nation, preservation in California communities serves as an object lesson for the rest of the nation, not only in land loss but also in redemption." Organizations like the Committee for Green Foothills and POST (the Peninsula Open Space Trust) have stemmed the rising tide of development to keep greenspace alive.

Consider Silicon Valley on a map. It's a nebulous designation with shifting borders roughly centered around San Jose. Highway 101 is its backbone, attached to an uneven ribcage of major freeways from roughly Palo Alto to South San Jose. This skeletal system supports a dense swath of urbanization. Despite the cityscape, however, if you were to stand on any high-rise with a 360-degree view in San Jose, you could easily observe the immense accomplishments of the open-space movement.

And what a view! "It is a view that has the quality of bigness without actual size, and it used to comfort me to know that these little mountains, like everything else around, are very lively, very Californian," wrote Wallace Stegner. The Santa Clara Valley, a.k.a. Silicon Valley, has always been "lively": even today it ranks as one of the most geologically active areas in North America, and its long tectonic history is full of booms and busts that can rival any tech-bubble cycle. The upheavals shaped this landscape of sublime extremes.

Foothills ripple across Silicon Valley on both sides, eventually cresting into two mountain ranges: Santa Cruz to the west and Diablo to the east. Open-space advocates have focused their efforts on preserving the foothills and mountains, even as a dense urban ocean laps at their feet. From an aesthetic standpoint, the greenbelt approach was successful. From an ecological standpoint, however, it left a fragmented ecosystem.

As we are finally learning, our favored approach to conservation—preserving islands of habitat—hasn't worked well. As Mary Ellen Hannibal states bluntly in her book *The Spine of the Continent: The Race to Save America's Last, Best Wilderness,* "Nature doesn't work without connection." Stand-alone protected areas are not sufficient—think about P-22 stranded in Griffith Park after his daredevil journey—and animals are disappearing even within national park boundaries, the best-protected places on the planet. Mary Ellen offers an example of this problem: Pronghorn use an ancient migration path in the Rocky Mountains that dates back more than six thousand years, and even though ranches and roads and oil drilling block their path, the animals refuse to deviate from their course. It is not enough to place them in a habitat of our choosing when instinct draws them elsewhere. Their insistence on following the path of their ancestors would have spelled their demise had it not been for a dedicated group of people—representing a range from industry interests to environmental activists—now actively working to help the pronghorn.

As for our foxes, they don't migrate, but they do need to disperse. Like P-22 and his mountain lion kin, foxes (and all wildlife, really) rely on genetic diversity to survive; isolation can lead to extinction, and city boundaries are formidable barriers.

Lucky for the Facebook foxes, they found an inviting habitat. As a result of a rising environmental consciousness, a trend of sustainable building, and the desire to keep talented high-tech workers happy, technology companies are now designing and building campuses that bring back greenery into the city. The campuses are not untouched wild areas, but foxes don't differentiate.

A gray fox named Cute, in Palo Alto.

When the foxes stood at the Facebook campus considering their next move, did they try to see past the concrete buildings and roadways? Did they sense the cattle grazing on the distant hillsides from 150 years ago? Could they smell the toxic metals swirling in the wastewater being pumped underground, or the lingering scent of apricots and cherry blossoms? Did they see the marshlands of 500 years ago, deafening with birdsong, or the flowing creeks lined with willow trees that their ancestors used to climb, now long cut down? When they marched onto the Facebook campus, and trotted past the first stunned employees, did the view of their historical birthplace persist, urging them on? Or did they simply do what worked, napping on the hood of a BMW in the employee parking lot, warmed by the sun, content with their new world as it was?

> "But there's one sound
> That no one knows:
> What does the fox say?"
> —Ylvis

The Norwegian musical comedy duo Ylvis's video for their hit song "The Fox (What Does the Fox Say)" went viral in 2013 and has almost five hundred million views to date. Their silly and clever electronic dance pop routine, featuring the artists dressed as the different animals along their onomatopoeic quest for a fox sound, stumbles—probably completely unbeknownst to them—on a legitimate gap in scientific inquiry. The truth is we really don't know much about what the gray fox says, or what he does, or how he behaves and relates to the world.

But the Fox Guy might have some answers. "I'm working on language now, and I think I am starting to understand their basic vocabulary," Bill Leikam tells me in our latest interview. He describes a recent encounter in the field, where he witnessed the arrival of a mate seemingly from out of nowhere, without any vocalization or other form of announcement from either animal, or at least not any he could detect. "I wonder how gray foxes communicate across distance without making a sound, or do they? Do they communicate as we normally understand that word, or is something else happening?" he wondered.

Bill's nickname is well earned. Having spent thousands of hours in the field observing *Urocyon cinereoargenteus* (the Latin name translates as "bushy-tailed dog of ashen silver"), he is the Jane Goodall of the gray fox world, immersing himself in the study of the urban foxes of Silicon Valley as fully as Jane did with the chimpanzees of Gombe, duplicating her assimilationist approach. He spends almost every day in the company of gray foxes. And the

foxes seem to enjoy his presence; he has a calm manner, a sooth-
ing voice, and a rhythm to his movements that put animals—and
people—at ease. A navy veteran, a teacher by trade, and a lifelong
citizen scientist, Bill has significantly advanced our knowledge of
the gray fox.

He entered the field of study quite by accident. One day while
walking in the Palo Alto Baylands Nature Preserve he stumbled
upon a fox sitting at the side of a dirt road. "I hadn't seen a gray
fox, much less thought of one, since I was about thirteen. The fox
at the Baylands just sat there as I took picture after picture, walking closer and closer, until I stepped around the edge of a gate. At
that moment, the fox stood, turned, and nonchalantly walked back
into the thicket," he remembers.

Something about the animal captured his imagination. "The
following morning I returned, but there was no fox. Two days
later as I passed through the area, I stopped and waited. From
out of the thicket emerged three young gray foxes. I was stunned
for I realized that I had come upon a family of foxes." The law
of the irresistibility of fox puppies applies as much to Bill as to
Mark Zuckerberg. He went back the next day, and the next. He
watched as the male fox streaked by, most mornings carrying
a field mouse or some other rodent to feed the pups. Though
gray foxes are omnivorous and will eat almost anything—insects,
birds, fruits, grasses—small mammals are a staple of a healthy
fox's diet.

Bill started making daily visits to the area, which he named Fox
Hollow, taking notes, posing research questions, and immersing
himself in the scholarship of these canids. "I enjoy seeing how
research questions unfold," he tells me. "They're always like picture
puzzles; the piece must fit perfectly before it's accepted."

That the puzzle of the gray fox hasn't yet been completely
assembled seems odd given both their wide distribution across
the United States and their status as the oldest living member of
the Canidae family. They represent a tie to the ancient world that

stretches back further in time than coyotes, wolves, and jackals—all animals we know significantly more about.

Fast-forward five years of sharing his days with the foxes and Bill was being recognized and lauded by scientists in the field. Dr. Ben Sacks, the director of the Canid Diversity and Conservation Unit at UC Davis, wrote a letter of commendation about Bill in which he said, "I was thrilled that he was studying the behavior of these foxes because so little was known about them....I have watched with great interest as Bill has sent periodic updates, photos, and film clips detailing his accumulated observations of the courtship, mating, pup-rearing, territoriality, and provisioning behavior of these foxes."

Bill's field reports sketch out the character of this enigmatic animal. Consider this excerpt from an entry titled "High-Speed Climbing School":

> [T]hrough the weeds, I saw a movement, and out from the weeds came Gray, the male of the "family," followed in a line—one right after the other—by his five pups. Before I realized it, the foxes were up in the lower branches of a huge eucalyptus tree. Down they came, out on a fallen branch from the tree, further out, out until the tip of the branch could no longer hold Gray and two pups, and they leapt to the ground. That was only the beginning of what turned out to be dad Gray teaching his pups high-speed tree climbing and navigation.

Surveying Bill's plentiful and exacting notes, you get to know the cast of fox characters. Gray, the attentive dad, is the mate of Bold, and they jointly reign over Fox Hollow. Bold, named for her tenaciousness, fought off her father, Squat, for possession of her natal den and territory. Usually the parents maintain the home turf and the youngsters disperse to new territory, but Bill is observing that in urban areas, where habitat is scarce, territorial boundaries

A gray fox pup giving its mother, Little One, a fox kiss.

are not as firm, and foxes bend the rules. He's documented two mating pairs and their offspring in close proximity, which would be unusual in wilder areas. He even witnessed the youngsters from both families playing together—again, not typical.

Joining Bill in his study of gray foxes and other urban wildlife in Silicon Valley is Greg Kerekez, who is also not a trained scientist. A wildlife conservation photographer and videographer, Greg grew up along the American River in Sacramento. "I spent my childhood fascinated by wildlife, collecting butterflies and frogs and other creatures all through high school to study them," he says. "When I found photography, I realized I could take pictures of animals instead of catch them."

For years, Bill had eschewed company on his excursions to see the foxes, and he initially resisted Greg's request to join him in the field for a day. But Greg's sincerity and persistence paid off, and after their first outing together, Bill sensed Greg's dedication to understanding the natural world and his genuine desire to help wildlife. He reconsidered his solo approach, thinking, "Maybe I am being too stodgy," and invited Greg out for a second and third excursion. This eventually led to their formation of the Urban Wildlife Research Project (UWRP), proving that opposites attract, even in the research world. A baby boomer, Bill retired from teaching in 2005 and tends toward introversion, while Greg's a millennial whose artistic expressions in film and music attest to his extroversion. Together, they are a formidable advocate for Silicon Valley wildlife.

In the field, they are two musicians playing a seamless duet. I joined them on fox patrol early one morning in the Baylands, watching the sunrise unmask a foggy landscape to reveal one of the largest tracts of undisturbed marshland in the San Francisco Bay. As we strolled toward Fox Hollow, blue herons flew by, and an array of ducks paddled along noisily in the slough. Some consider the Palo Alto Baylands the best place for bird watching on the West Coast, and for the same reason that foxes like the area.

As we waited for the foxes to appear, Greg gave us his best fox impression—a hoarse call that sounds like a mix of a raven's caw

A fox kiss greeting.

and a yippy dog's bark, each with a case of laryngitis. In contrast to their wolf and coyote relatives, gray foxes don't howl and are not rated high as songsters. Not surprisingly, they are at their most communicative during mating season—a time that brings out the chattiness of many in the animal world—and their repertoire also includes mews, coos, and some growls and snarls. Yet, vocalization may not be the most important form of communication for this canid. The gray fox employs subtle tactile, chemical, and visual signals in social encounters, as Bill's diligent and constant observation has often uniquely documented.

As if the foxes could get any more endearing, their traditional way of welcoming another fox is with a kiss. And these are no mere cursory pecks on the cheeks. Pups will eagerly touch noses with their parents, lowering themselves to the ground and curling under the adult's chin, while males and females who have been separated will swish their tails and snuggle into each other, celebrating their reunion by rubbing snouts. Under certain conditions, adult foxes from different families may use the affectionate greeting as a sort of handshake.

But it's not all kisses and group hugs in Fox Hollow. Gray foxes have their conflicts over territory and food, like any other animal, but they lack the violent and lethal outcomes typical of wolves, who will often kill on sight other members of a rival pack. Foxes even have their own "peace sign" to avoid fights, which "works about 97 percent of the time," according to Bill. To placate another fox, the animal will approach with its belly low to the ground, swish its tail, and sometimes give a fox kiss, maybe rolling on its back and speaking in a series of high-pitched squeaks.

Foxes are largely amicable, engaging in mutually beneficial practices such as grooming and other displays of kinship, including this one Bill and Greg documented in their research: An adult female appeared one day in the Baylands with no mate or pups. She wandered into the territory of foxes named Creek and Little One, who had a den site with four pups alongside the levee road near Matadero Creek. As the family considered her, they recognized her as Little One's sister. They made no move to chase her away as they typically would an ordinary trespasser. This new female was the first documented instance of helper females within the gray fox community. She helped raise the pups. She hunted for them and brought in food, and she played with them, teaching them the basics of being a gray fox. She became a fully accepted member of the family, and for the entire childrearing season she stayed and helped. Then, even after the pups were grown and had dispersed, this wolf, whom Bill named "Helper," remained in the vicinity.

As Greg remembers, "We were stunned. This provided the first documented occurrence of a helper female in the gray fox world." And Helper didn't just make history, she also instigated a soap opera. "Interestingly enough, the male 'Creek' disappeared the next season, and the two adult females 'Helper' and 'Little One' stayed together, neither of them having pups," read Bill's report. The following season, Helper and Little One remained on their own until another male, Brownie, arrived, who whisked them both away to another location. Foxes are monogamous—to a point—although this example shows that foxes in urban areas with less habitat may prove less faithful than those that live in wilder areas.

While we walk, I notice that this 1,940-acre nature preserve is still a highly peopled landscape. We pass visitors hiking, running, biking, walking their dogs, bird watching, and canoeing. The city's airport and the landfill border the park, and some of the foxes have made their dens in close proximity to the nearby water treatment plant. It now made sense why the Facebook foxes, who likely traveled to the campus from the Baylands, didn't shy away from denning in the highly trafficked Zen Garden or napping on cars in the staff parking lot.

Bill makes a distinction between the urban gray fox and its wilder counterparts, citing behavioral differences that come with city living: "These urban gray foxes have frequent interactions with people. As such, they do not fear being around people, but at the same time they will dash away if a person approaches too closely. Unique to the urban fox, adult gray foxes seem to intentionally introduce their young to people, as seen by direct observation. This is in contrast to the gray foxes that, for instance, live in the wilderness of the nearby Santa Cruz Mountains. These gray foxes are seldom seen."

We created the new world of Silicon Valley, but the urban foxes might be guiding us in how to live there. In their folklore, many of Northern California's Native American tribes consider Silver Gray Fox a cultural hero—he creates the world and teaches people how to live in it. I see the Facebook foxes as emissaries, and as they do with strangers of their own kind, they are making those first tentative peace offerings toward us in hopes we'll share our territory. Foxes can adapt to us. It remains to be seen if we can adapt to foxes.

"I'm a Silicon Valley guy. I just think people from Silicon Valley can do anything."—entrepreneur Elon Musk

"Carnivores, fortunately, are here to stay, and so are cities. More than 50 percent of the world's human population now resides in urban areas....A bobcat in Tucson, Arizona, a coyote in Los Angeles, California, a raccoon in Chicago, Illinois, and a red fox in London, England, can all add significantly to people's experience of nature, benefiting the people, the animals, and we hope, in the long run, the wider world."
—*Urban Carnivores: Ecology, Conflict, and Conservation*

Wildlife are returning to the Santa Clara Valley. Although the landscape will no longer support abundant herds of tule elk, the pronghorn have long vanished, and the closest grizzly bears are in Washington State, we are not without hope. We may never return to the past of plentitude, yet something new is emerging—a model of urban refugia. Our actions, while not perfect, seem to have issued a clarion call for wildlife, and animals, like the Facebook foxes, are tentatively answering.

After more than 150 years, beavers returned to the Guadalupe River in the spring of 2013, busily gnawing trees near the Shark Tank (the SAP Center) in the downtown river parkway in San Jose. Greg discovered the beavers, first an individual and then a family, which he features in his "Beavertown San Jose" video series. In 2013 scientists also documented the first successful nesting of Swainson's hawks in Santa Clara Valley since 1894, and noted in their research paper that conservation efforts might have played a role, "or that it [the hawk] may be adapting to human-modified habitats." The endangered condor returned to Silicon Valley in 2011, when five of them were sighted soaring over Mt. Hamilton—not twenty miles from San Jose—and then hanging out not on a ponderosa pine or blue oak but on the roof of the Lick Observatory dome.

All of this echoes the situation several miles north in the bay, where harbor porpoises have returned after a sixty-five-year hiatus. As Bill observes, "I think we are seeing that some of the original ancient passageways are partially intact and that the animals in some cases are not giving up, even with cities moving into their areas."

Yes, wildlife would be better served in a Silicon Valley of total open space, undeveloped and restored to the native landscape. But barring a stunning reversal of the trends of human existence, this won't happen any time soon, and a developed Silicon Valley that incorporates wildlife is surely better than one that doesn't. People should continue advocating for protections and restorations, knowing that the wildlife already here is probably better served by a more porous view of nature and cities.

Wildlife are certainly adapting and giving it their best efforts. On an airport landing strip in San Jose, two burrowing owls hatched to different families played in the fenced-in fields and became used to the deafening noise of jet planes. After they grew up, they decided to pair up, at which point they journeyed away from the strip, hoping to find a patch of land they could call their own. They settled on some of the only open space available to them, right next to a sidewalk and road. The city intruded constantly, automobile noise never ended, helicopters flew overhead frequently, campers parked on the road nearby, and people walked their dogs off leash. And still the owls prevailed on their small plot of green. Every day at lunch, people would stroll or bicycle by and marvel over the owls, the last outpost in the city. Soon the female hatched six chicks, which also drew crowds. One day, the male owl disappeared, probably a victim of urban-caused death: hit by a car, attacked by a dog or cat, or poisoned by pest control bait. The female soldiered on, miraculously managing to raise all six chicks herself. She hunted constantly to forage enough food for them, running herself ragged. Soon after her chicks had grown, she heard the loud sound of machinery as the bulldozers approached.

Unlike many of the stories I relate, this one doesn't have a happy ending. This owl's home, Orchard Parkway, is now plowed over, not even a small green patch left. The burrowing owl, once ranked as one of the most common birds in California, has in the last two decades lost 60 percent of its breeding pairs. In the Santa Clara Valley, predictions give them less than two decades before they disappear entirely from the area. The burrowing owls of Orchard Parkway, like the Facebook foxes, tried to adapt to the urban Silicon Valley lifestyle. Yet unlike the foxes, they didn't stumble on a campus with a Zuckerberg-type leader and twenty-five hundred employees championing their survival.

In Silicon Valley, however, there remains some hope that the high-tech industry will be good for the local wildlife. A generation of high-tech workers are coming to expect the natural world as an integral part of their work environments. Facebook's new campus, just across the freeway from their old one, is designed to "naturally fit into the surrounding marshlands," and it offers a green roof with transplanted trees from across California as well as drought-resistant grasses. Architect Norman Foster described his plans for Apple's new campus in Cupertino as "essentially a park that would replicate the original California landscape." It will cover only 13 percent of its 176-acre site, and six thousand trees will cover the landscape that also includes 15 acres of native grassland. The complex's circular structure and attention to ecological planning almost gives the impression that Apple is attempting to build its own bio-dome.

A technology campus might not be the traditional view of nature for most people, yet these sites are paving the way for the future; they are taking their landscape into account, in some cases partially restoring them, or perhaps the right term is "reenvisioning." The San Francisco Bay little resembles its self of two hundred years ago, yet the porpoises still approve. The Facebook campus—and all the

A mother burrowing owl feeding her chick.

new nature-inspired campuses—might not be the Santa Clara Valley of the past, but the foxes had no problem setting up house.

And this new nature, taken in totality, just might lead to something pretty significant for conservation. Bill and Greg have a vision. Beyond studying the behavior of the foxes, they are also charting something fundamental and vital to the future of foxes and all wildlife in Silicon Valley: how to maintain and reestablish ancient and modern pathways through the land. Bill and Greg's dream is to establish a San Francisco Bay Wildlife Connection Corridor. Utilizing public and private lands, and enlisting the cooperation of a wide array of government agencies, nonprofits, businesses, and individuals, they hope to restore enough connectivity in Silicon Valley to create a thoroughfare that allows animals passage as far north as Oakland and San Francisco. It's as ambitious as it is necessary if wildlife is to have a future here.

Can we unite the high-tech companies, the cities they are located in, and the residents who live there to develop the version 2.0 of wildlife refuges, to create something innovative like a High-Tech Open Space District? Can we open-source open space? These companies and their leaders have transformed so many lives with technology, and they can do the same for urban wildlife. What if Apple, Google, Facebook, LinkedIn, Adobe—all the titans of Silicon Valley—all had Certified Wildlife Habitat campuses that were connected into this San Francisco Bay Wildlife Connection Corridor? What if the neighborhoods surrounding the high-tech campuses took the lead and certified their yards? What if Palo Alto or San Jose or Mountain View united to create a corridor that foxes and other wildlife can travel through unimpaired?

Silicon Valley's wildlife have decided to give us a second chance. They forgave us cattle grazing, orchards, development burying the marshes and wiping out grasslands, toxic soups and litter choking their waterways, asphalt and automobiles blocking their pathways. They can adapt to urbanized spaces, but not to total oblivion. Facebook has this motto painted on one wall: "Done is better than perfect." Let's apply this to our efforts to make room for wildlife in urban spaces. Burrowing owls, foxes, beavers, and their kin are willing to compromise. Are we?

Bakersfield Shares a City with Endangered Foxes

More than ten thousand San Joaquin kit foxes once lived in California's Central Valley, but today the endangered animal struggles to survive, as its traditional habitat has largely vanished. One population of kit foxes has been successful at making a new home in the most unlikely of places: Bakersfield, the ninth-largest city in the state.

Brian Cypher, a biologist at Cal State Stanislaus, commented on the foxes adapting to life in the big city to the *LA Times*: "At first, we thought they were displaced stragglers that would be pushed out or die off as development continued. But they're doing surprisingly well in the urban area." An estimated four hundred foxes live within the city limits, making their dens in such unlikely places as schoolyards, athletic fields, culverts, and golf courses. Dens are key to these creatures, who occupy their homes year-round to escape the extreme heat. Some groups have even installed artificial dens for the foxes, such as the Seven Oaks Country Club, which constructed one right on the green and has seen generations of fox families raised on its golf course.

Don Ciota, general manager of Seven Oaks, says his organization is committed to sharing the space. "We respect we are on their habitat," he says. "Members like having the foxes around and we all try to live in harmony." He points out that life with the foxes, however, is not without its challenges, citing their habit of stealing food from the clubhouse or golf balls during games; a remote camera revealed a stash of hundreds of balls in one den. Mischief aside, the foxes seem to be accepted members of the club, which even has a local kit fox rule on the scorecard.

A San Joaquin kit fox at Seven Oaks Country Club.

Peregrine Falcons Nesting at San Jose City Hall

On my visit to San Jose City Hall, Michelle McGurk took me to the mayor's office and introduced me to the city's most famous resident, Fernando El Cohete, who stared at me intently through a window. "Fernando the Rocket" isn't the mayor but one of the peregrine falcons that nests and hangs out on the ledges of the tall building.

Michelle walked me over to view the nesting site and pointed out a bird carcass on the windowsill. "City employees are used to seeing the remains of the meals on windowsills," she laughed. "We like sharing space with [the birds]." And it's not just the staff that like having them around. These winged neighbors proved so popular that the City of San Jose and the Santa Cruz Predatory Bird Research Group at UCSC set up the online FalconCam, which attracts viewers from around the globe. The falcons have called City Hall their home since the matriarch, Clara, appeared in 2006. She has nested there every year since 2007.

A skyscraper in the middle of a large city might seem an odd choice for a peregrine home, but as Glenn Stewart, director for UC Santa Cruz's Predatory Bird Research Group, points out, "Urban areas have a plentiful food source—pigeons—and the structures act as artificial cliffs as peregrines need a launching place to hunt." They are the fastest animal on the planet and dive at speeds of more than two hundred miles an hour. "What they do would rip the wings off an F-15," Glenn remarks.

Glenn has studied the bird since the 1970s and was part of the team of scientists that helped them recover from the brink of extinction. "In 1970, we could only find two pairs of peregrines in the entire state," he says. "Today we have about three hundred pairs. They are a great Endangered Species Act success story."

Glenn Stewart of the Santa Cruz Predatory Bird Research Group banding a young peregrine falcon on San Jose's City Hall.

Google Provides a Safe Haven for Egrets

Facebook has its foxes, but it's not the only tech campus with an accidental mascot. The Google campus, or Googleplex, in Mountain View houses one of the largest egret rookeries in the Bay Area, and the company is taking steps to ensure its avian tenants remain.

As Google spokeswoman Meghan Casserly told the *San Jose Mercury News* in 2015, "Mountain View is our hometown and it's important to us to be respectful of our neighbors, whether they walk, drive, burrow, or fly. We're happy to make accommodations for nesting season by keeping traffic closed on Shorebird [Way], the volume down on that part of campus, and also by educating Googlers on how to appreciate the egrets' beauty from a safe distance."

In 2014 researchers estimated the rookery held forty-one snowy egret and thirty-six great egret nests. Along with taking special precautions during nesting season, the company has considered the birds' needs during the campus' expansion project by designing spaces that respect their flight paths and proposing the removal of existing buildings to provide better connectivity to the wetlands surrounding the campus. Google also supports research, care, and monitoring of the birds and actively works with the City of Mountain View, the Santa Clara Valley Audubon Society, the Wildlife Center of Silicon Valley, International Bird Rescue, and the San Francisco Bird Observatory.

The arrival of the graceful egrets every spring makes for a beautiful—and loud—spectacle on campus. As Shani Kleinhaus reports in the *Avocet*, "Every year, dozens of egrets return to nine trees on Shorebird Way to reconstruct nests, mate, argue with neighbors over choice twigs and branches, and, eventually, raise a family in a noisy cacophony."

A great egret near the Google campus.

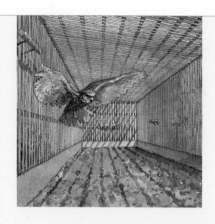

Larry Ellison Founds an Innovative Wildlife Center

As first reported in the press in 2015, Oracle's Larry Ellison is funding development of a next-generation center dedicated to native wildlife. Located in Silicon Valley, the Innovative Wildlife Center will be operated as a nonprofit organization. I spoke to Ken White, president of the Peninsula Humane Society and SPCA, who is responsible for getting the center built.

How did Larry get involved in this project? Every humane society is independent—not chapters of the groups with national-sounding names—and the Peninsula Humane Society is Mr. Ellison's local animal welfare organization. We've a broader mission than most, as along with handling homeless dogs and cats, we also care for injured and orphaned native wildlife. Frankly, it isn't easy to attract funds for squirrels, opossums, snakes, and other less charismatic animals. One day, Mr. Ellison asked: What would it take to address wildlife's needs locally? This center is the answer. We're not going to save the world, but this project will do a lot for this corner of it, and we hope other communities will replicate.

What are some of the wildlife you'll be helping? We anticipate ten thousand animals annually at a wildlife hospital unlike anything previously imagined, including pelicans, murres, hawks and owls, songbirds, deer, raccoons, foxes, and squirrels. But in addition to helping orphaned babies and injured adults, perhaps the most exciting to me is our partnering with state and federal wildlife agencies to breed rare and endangered species of invertebrates, reptiles, and amphibians—like the San Francisco garter snake or mission blue butterflies—for release to the wild. It's wonderful that people are working to save condors or tigers, but the truth is we can't function without, as [biologist] E. O. Wilson calls them, the smallest building blocks of nature.

Being located in the middle of Silicon Valley, the new center will likely use leading technology. Absolutely. We're talking about utilizing things like DNA fingerprinting for monitoring. But in reality, when you are dealing with a raccoon with diarrhea, there's no app for that one.

This page: A design for a large raptor aviary. Opposite page: Yellow-faced bumblebee buzz pollination.

Farming for Native Bees

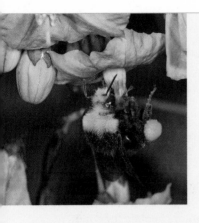

Next time you see a patch of flowers humming with bumblebees, listen for what entomologist Stephen Buchmann describes as a sound like a bee "giving you a raspberry." It results from "buzz pollination," their special method for retrieving pollen that is akin to us shaking apples from a tree—if we could perform this while hovering midair and then catch the apples electrostatically with our body hair.

Bees pollinate an estimated 75 percent of the fruits, nuts, and vegetables grown in the United States and agricultural products worth $15 billion annually across North America. Large-scale pollination is chiefly driven by human-managed honeybees, an introduced species to the United States, yet their recent widespread declines have shifted the focus away from our dependency on this single species for food production. More than fifteen hundred species of native bees live amongst us in this country—40 percent of whose range is at least partly in California.

"Making small changes to increase the number of native pollinators on a farm does not require a lot of work," advised the Xerces Society for Invertebrate Conservation in a recent report. The organization implemented a number of collaborative programs with agricultural interests, such as their partnership with Muir Glen Organic Tomatoes, which created a pollinator corridor of a mile-long native plant hedgerow. The UC Berkeley Urban Bee Lab, after conducting twelve years of vanguard research on urban bees, formed its Farming for Native Bees initiative, with projects such as working in Brentwood with eight small farmers, including the popular Frog Hollow Farm, to increase native bee habitat. "We're identifying the key native species that move in between crops and other plants, and focusing on attracting them by enhancing habitat. It makes a big difference in crop production," notes project manager Sara Leon Guerrero.

But you don't need to be a farmer to help with native bees. As Peter Bernhardt points out in his foreword to *California Bees and Blooms,* bees are "easily coaxed back into urban areas." To encourage all citizens to help, in 2015 the National Wildlife Federation joined with dozens of conservation and gardening organizations to form the National Pollinator Garden Network and launched the Million Pollinator Garden Challenge.

Elephant Seals: The Comeback Kids

I am helplessly in love with elephant seals. I never tire of watching their antics, which range from those of an affable dog to a curmudgeonly grandfather to a boastful prizefighter. They sound constantly annoyed with their barks and yells, yet enjoy cuddling together while napping on the beach. For some people their incomplete elephant's trunk of a nose is a barrier to admiration, but I find it ridiculously cute-ugly. Jacques Cousteau, in his 1974 book *Diving Companions,* wrote about his affinity for the creature: "At the beginning, the animal seemed unattractive, even repulsive....Yet, we ended by experiencing a real sympathy for these giants who are the victims of their own size. We discovered their virtues, their courage and tenderness, and love of freedom."

Independent and intrepid by any measure, elephant seals spend most of their lives diving at sea, resting only for minutes at a time at the surface in between plunges of up to a mile below. Marc Webber, coauthor of *Marine Mammals of the World,* related to me that recent research shows that despite their bulk of almost a ton, "elephant seals dive by slowly descending through the water, like a falling leaf spiraling to the ground, and probably while sleeping."

California hosts unparalleled elephant seal viewing. Año Nuevo State Park, one of the largest mainland breeding colonies in the world, provides guided walks on the beach, while Piedras Blancas in San Simeon has public viewing platforms at a rookery right off the highway. That northern elephant seals survive on the California coast at all attests to the fortitude of the animal. By the late 1800s, they had been hunted almost to extinction, with less than a hundred surviving. After Mexico and then the United States enacted protections, the population rebounded, and today more than one hundred thousand of these animals visit California's shores.

An elephant seal at Año Nuevo State Park.

Up Close with a California Condor

The first time I saw a California condor in the wild, I was releasing it from a crate, returning the bird home after scientists had treated it for lead poisoning. A condor in flight is impressive, and all of us—even the biologists who work with them daily—stood silent and awestruck as this ancient giant effortlessly glided over the hills, the wind playing its long wings like a musical instrument. They can travel more than two hundred miles in a single day and soar fifteen thousand feet above the earth. Some Native American tribes attributed the thunder to their flight.

That day in 2014 I had accompanied a team of scientists working on condor recovery at the Hopper Mountain National Wildlife Refuge, led by Joseph Brandt of the US Fish and Wildlife Service (follow their Facebook page "The Condor Cave"). Sadly, all but one of the thirteen condors sampled from the wild that we examined that day tested positive for lead exposure.

Scientists captured the last wild condor in 1987 and then began an extensive captive-breeding program to help save the species. Reintroduction to the wild began in 1992. Since then the bird has started expanding back into its historic range—the first wild condor in more than a hundred years was spotted in San Mateo County in 2014—yet its future remains uncertain. As Joseph points out, "Without captive breeding, studies have shown it would take over one thousand years for the bird to fully recover, in spite of other intensive management, because of lead exposure and other human threats. If we also stopped treating condors for lead exposure, the population would return to the brink of extinction in a single generation of condors." In 2013, California took the unprecedented step of banning lead bullets for hunting; after a phasing-in period, the law takes full effect in 2019.

A California condor at the Hopper Mountain National Wildlife Refuge.

Sequoia, a resident gray wolf at the California Wolf Center.

CHAPTER 5

The Incredible Journey

CALIFORNIA WELCOMES BACK WOLVES AFTER NINETY YEARS

What does California remember of wolves?

Do the tule elk, those scattered remnants of the wild herds that once roamed California in the hundreds of thousands, recall the wolf packs that gave chase across the wide grasslands of the Central Valley? When the elk trot stiffly, arching their necks and holding their heads high, are they putting on a confident display for a predator long gone? Do they still have a warning call reserved for wolves, unused for generations? Would they shiver if they heard a howl carried over the hills by the wind? Or have even the elk forgotten that haunting song?

Does the condor, itself nearly vanished into legend, imagine wolves loping on the hillsides as it soars overhead, its magnificent wings casting moving shadows on the land like clouds do? As it glides on the rising thermals of air scouting for carrion, does it remember the bounty wolves brought to its kind, when it could follow the hunt from above, watching the wolves conquer an elk or deer and be assured of the leftovers? Does each generation of condor pass along renewed hope of finding a wolf in their travels?

Do the coyotes celebrate the banishment of their onetime nemesis? The absence of wolves to keep them in check has assured their recent reign as the dominant canine in all corners of California. Does the coyote remember a time when it had to abandon a carcass when wolves appeared, dashing away at full speed, because the price of a meal in wolf territory could mean death?

Do the timid kit foxes, suffering from the unrestricted harassment of the coyote, long for a day when the wolves might return to end the unbalanced regime? Do the wise and long-lived ravens, as they feast on a roadkill brush rabbit or the half-eaten remains of a hamburger bun, remember a time of plenty when they led wolves to the elk herds and were rewarded with rich scraps from their kills?

And the willows and cottonwoods and oaks snuggled by the riverbeds, and the grasses and wildflowers coloring the meadows in spring, and the blackbird feasting at the elderberry tree, and the red-legged frog resting in a vernal pool—do they retain a collective memory of the almost-forgotten world once shaped by wolves?

Do wolves remember California?

Do they remember the bellowing of the tule elk resonating across an almost limitless playground of the San Joaquin Valley, where they could lope for miles over the rippled hills and rest in the shade of the riparian oak woodlands? Do they remember hunting under the watch of the tall redwood forests, or splashing about in the marshes near the shores of the San Francisco Bay, relishing the abundance of prey in this bountiful land, thick with herds of pronghorn, elk, and deer? Do they remember the moonlight glinting on the polished granite of Sierra Nevada peaks, or having to relinquish the hard-earned kill of a mule deer to a grizzly in a mountain forest? Do they remember the dense, salty smell of the ocean or the sharp, arid air of the desert?

Whether the wolves have forgotten the scent of the Golden State or the condors and coyotes and elk of California have forgotten the music of the wolf, it doesn't matter. A landscape is regaining its memory.

The wolf has returned.

"Wolves may feature in our myths, our history, and our dreams, but they have their own future, their own loves, their own dreams to fulfill."
—songwriter Anthony Miles

Just a few days after Christmas in 2011, a lone gray wolf wanders in a ponderosa pine forest sniffing for hints of a meal after having traveled more than a thousand miles across two states. He paws the frost-hardened ground to scavenge. Not picky after such a long road trip, he'd be happy with even the meager fare of a hapless rodent. In between his searches for food, he raises his head to give a lonely howl.

As he trots a haphazard route southward in his explorations, he knows nothing of state lines, or of the GPS unit in the collar he wears that tracks his footsteps. He doesn't know about the millions of people around the world avidly following his journey, or that their eagerness for news of him will crash the California Department of Fish and Wildlife's website. He's ignorant of the front-page headlines and the thousands of tweets and Facebook posts that trend on social media about his historic trek.

He didn't know when he awakened that morning after resting on the snow-covered ground, head curled under his tail to conserve warmth, that he is the most famous wolf in the world or that he has become a cause célèbre for the future of wolf conservation. He doesn't know that later that day he will become the first wild *Canis lupus* to set foot in California in almost ninety years.

What he knows is hunger. The need for a home unoccupied by a rival pack. And the desire for a mate.

In retrospect, OR-7, as this wolf is known, seemed destined for greatness. He comes from pioneering stock, direct descendants of the first wolves introduced into Yellowstone and central Idaho in 1995 as part of one of the most significant conservation efforts of our time. His mother, an impressive gray wolf known as Sophie (or OR-2 or B-300), also deserves a mention in the history books. She swam across the Snake River in early 2008 to cross into Oregon from Idaho; while wolves can swim aided by the webbing between their toes, Sophie certainly risked drowning in attempting the icy and swift Snake. After meeting up with OR-4—at 115 pounds, the biggest wolf recorded to date in the Beaver State— they became the Adam and Eve of Oregon's wolves, forming the first wolf pack there in more than sixty years. Their second litter included OR-7, and he inherited their wandering ways.

What compelled him to leave family behind, led him so far from his known world, drove him hundreds of miles into territory where long ago humans had banished wolves? Did he trot across the state line in the true fever of exploration, ignorant of what lay ahead, but eager to find out? The nature of wolves is to disperse

A remote camera photo of pioneering wolf OR-7 in 2014.

A map of OR-7's travels in California.

far and wide, yet it's a lot to take on faith that continuing blindly south would yield a home; even Lewis and Clark knew they would get to the West Coast eventually.

Did he have some sort of guidance? Did he study a historical geography of wolves that named California as a potential refuge? If a map designed by wolves does exist, it's one based on smell—contour lines replaced with shifting boundaries of odors, routes marked with the perfume tracks of wild animals, the fragrance of rock, the aroma of a forest. Wolves possess more than two hundred million scent cells, as compared with the mere five million housed in our human noses. Even more remarkable, scent for wolves extends through both space and time. As Yellowstone biologist James Halfpenny reminds us, "Do not underestimate the power of wolf odor communications." As OR-7 trotted through hill and dale, he navigated with his nose. Scent has memory for wolves. By sniffing the air from more than a mile away, he could determine if wolves lived nearby, whether they were male or female, young or old, their mating status and rank in pack hierarchy, and even how long ago they had passed.

Even with this extraordinary talent, however, the trip to California wasn't an easy one. Most wolves disperse no farther than sixty miles from their original established turf, yet OR-7's wanderlust propelled him more than five hundred miles as the crow flies from his birthplace in Oregon. Despite the persistence of the lone-wolf mythology, it's rare for wolves to set out on their own—a mere 15 percent risk going solo and leaving the relative safety of the pack. Mortality for these loners trends much higher than other wolves, as life on the road as a single wolf can be perilous; they are more easily targeted by rival wolf packs, and bringing down larger prey like elk and deer is tough going without a team. Other threats, like the challenge of crossing roads (one lone female wolf traveled from Yellowstone to Denver only to be hit by a car) or being killed by humans (a poacher illegally killed OR-7's brother OR-9 in 2012), make a lone wolf's odds of survival slim.

OR-7 kept defying the odds. He passed many wild places of suitable wolf habitat, and even sauntered through territory occupied by fledging packs that he might have attempted to join. Yet he kept going into the unknown.

While dining on mostly mule deer, at least one elk, and not a few squirrels, he took in the sights, with stops at the usual tourist destinations, such as Crater Lake and Mt. Shasta. He roamed close to the city of Ashland and checked out Lava Beds National Monument. For the most part he avoided populated areas and wandered through some of the wildest places in the Pacific West.

And still he kept going, edging closer to California, at this point watched by millions of online followers across the globe who eagerly tracked the colored lines that traced his route on each newly updated map. (The updates were time delayed for OR-7's safety.) Would he make history? Finally, on December 28, 2011, he trotted across the Oregon border into the Golden State.

Whether OR-7 wandered into California out of instinct, igno-
rance, or an urge to "light out for the territory ahead of the rest"
and perhaps find a new wolf Shangri-La, his journey captured the
imaginations of people around the world and, for Californians,
signified something even more: a sense of what journalist Matt
Weiser called "wild wonder." As Matt reported, OR-7 made us
feel that we were "part of something miraculous. No matter our
politics or lifestyle, we were astounded that our crowded state
could still interest a wolf."

"California has more people with more opinions than any other state. We have people calling, saying we should find him a girlfriend as soon as possible and let them settle down. Some people say we should clear humans out of parts of the state and make a wolf sanctuary."—Mark Stopher, senior policy advisor, California Department of Fish and Wildlife

"OR-7 is a family wolf now. And, like a lot of Californians, he went to Oregon to raise his family."—Chris Roberts, *NBC Bay Area*

"Come back #OR7. Bring pups! Wolves just granted endangered species protection by California Fish and Game Commission."
—@the_wrangler on Twitter

After venturing into California, OR-7's celebrity stock rose considerably. Followers deemed the classic rock group Journey's "Don't Stop Believing" his official ballad, and when the conservation group Oregon Wild sponsored a children's contest to name him, online voters ultimately picked, among entries from around the globe, "Journey," submitted by two children, ages 7 and 11, from Idaho and North Dakota. Although this Hollywood name hasn't quite caught on in California, our star did land a movie deal (the film *OR-7: The Journey* was released in 2014—see page 170), he became the subject of a musical dance production (one article advertised it as "Journey: The Ecokinesis Dance Company and Crane Culture Theater tell the remarkable story of the California wolf"), and he of course acquired his own Twitter account and Facebook page.

After stepping across the state line in December 2011, OR-7 continued his vagabond ways, remaining exclusively in California for three months before returning to Oregon for brief visits in March of 2011. He spent most of 2012 wandering in a zigzag fashion through Northern California, bouncing roughly between the I-5 and 395 highway corridors, rambling almost as far south as Chico. For a few weeks during this time, he roamed around the perimeter of one of the large wildfires burning in the state, causing some biologists to speculate that he was waiting for prey fleeing the flames. In April of 2013, he reentered Oregon, hung out west of the Klamath Falls area, and popped over the state line to California only twice the rest of the year.

Despite his extraordinary journey, most assumed OR-7 had embarked on a futile quest for a mate and they predicted he would die a lonely bachelor or admit defeat and head north back to known territory. Scientists considered his presence so far south an abnormality; OR-7 might be the harbinger of California's wolf future, they agreed, but he was ahead of his time. Almost every article written about him lamented the unlikeliness of a mate being within howling distance, and the public expressed sympathy for his plight, assuming the nearest female wolf was probably half a state away at best.

Even researchers had given up on OR-7, at least in terms of his ongoing scientific value. With the batteries on his tracking collar running low, biologists debated the usefulness of recapturing him and replacing it. He had made his name in history by crossing into the Golden State, but what more could he contribute? In typical California fashion, after his fifteen minutes of fame, he seemed destined to return to a private life, crossing state lines without his public watching. The reality show of OR-7 was about to come to an end.

In March of 2014, however, with his GPS collar still functioning, signals indicated that he suddenly appeared to have had enough of the road. He was getting pretty old for a wolf—at five years old, he'd survived past the average age of most of his kind. His sudden slowdown intrigued the biologists who monitored his behavior, as this new trend coincided with mating season. Staying put usually indicates a courtship in bloom in the manner of wolves. As John Stephenson of the USFWS recalled in a *National Geographic* article, "We were curious. But we always felt it was a long shot that a female would find him."

In May of 2014, photos from the camera traps placed near his newly adopted territory just north of the California border revealed a black-coated female wolf squatting to urinate as she stared down the camera. The discovery of his mystery mate excited a fresh wave of tabloid-like media frenzy and had California's wolf-recovery advocates giddy at the news. The wolf was literally at the door.

Where did she come from? She wasn't a known wolf, and speculation ran high. DNA collected from scat samples later revealed her relationship to wolves in northeastern Oregon—OR-7's hometown. OR-7 had met the girl next door five hundred miles away.

OR-7 humbled us—his imagination surpassed ours. And his meeting one of his kind this far from home, finally hearing an answer to his lonesome howls after almost three years, spoke to the beauty and resiliency of wild things, one that most of us sense only from the margins. Through his story, outsiders comfortable in the security of the modern world could receive a glimpse of how all the human truths we cling to don't impress a wolf. Wolves don't operate by our rules and laws, and their ability to defy our expectations is something to celebrate, along with their ability to maintain some secrets, such as keeping a litter of pups under wraps. Yes, not long after the first photo of OR-7's mate surfaced, it became evident that they'd been courting on the sly for some time.

You don't often see birth announcements for wolves, yet when a camera revealed that OR-7 had become a proud papa, the ridiculously cute wolf puppy photos released by the Oregon Department of Fish and Wildlife attracted worldwide attention. The headlines were nothing less than celebratory. "OR-7 Now a Dad!" "Adorable Animal Babies: Meet Wandering Wolf's First

Pups from OR-7's history-making first litter pose for the camera in July 2014.

Pups." "Happy Father's Day to OR-7!" One story led with this
line: "After just three weeks of simmering will-they-won't-they
tension, wildlife officials report that OR-7's mate has given birth."

If you watch a wolf pack regard its pups, you'll recognize some
very human expressions, ranging from pride to good-natured
annoyance as the pups take their first steps or tug and bite ears
and tails. Beyond affection, the devotion wolves display toward
their young is rooted in maintaining the strength of their dynasty.
Rudyard Kipling captured the mantra of *Canis lupus* perfectly
when he wrote in *The Jungle Book,* "The strength of the pack is
the wolf. And the strength of the wolf is the pack." The energy
of the entire pack is dedicated to the survival of the pups, and all
members, whether related to the new litter or not, lend a hand
in their care and feeding. Nature even provides a failsafe guarantee: when a litter arrives, the hormone prolactin, associated with
nursing, surges through all members (even males) and fills everyone with familial goodwill.

Lacking the support of other wolves during their first year together, OR-7 and his mate didn't have much of a break from parenting. But the family has grown with the addition of a second litter, in 2015, and the group is now officially recognized as the Rogue Pack. (Pack status is usually obtained after four or more wolves survive a winter together.) Pack size can vary considerably, usually due to the availability of food and open territory—the largest wolf pack recorded in recent history, the Druids of Yellowstone National Park, peaked at thirty-seven members in 2001—but whatever the size, this cohesive family unit possesses an intricate structure, not unlike our own. In his *Wolf Almanac,* Robert H. Busch quotes Stephen Young, founder of the Center for Northern Studies: "Wolves are considered by many to have the most complex social behavior of any non-primate animal."

At the center of this structure is the breeding pair, the term preferred by wolf expert David Mech in place of "alpha," as studies have shown leadership roles can shift within a pack—that is, there is no fixed top dog. Typically, a pack supports only one breeding pair in order to avoid inbreeding, so wolves need to disperse to find a mate, sometimes seeking independence as young as ten months, although most wait until they are older.

Aside from catapulting a wolf and his progeny to fame, the media attention brought by OR-7 helped shape the wolf recovery debate in California and the Pacific West. The photos of two of OR-7's first puppies peeking out from behind a log became "a small, adorable conservation success story," as writer Jason Goldman proclaimed. As the youngsters barked at squirrels or played tag with each other, the California Fish and Game Commission met two hundred miles south of the den site and voted to do something absolutely unprecedented. They used a state act to protect an animal that hadn't even set up permanent residence in California.

Photos of OR-7's pups—now big-pawed, awkward adolescents—flashed on the screen in June of 2014 during a meeting in Fortuna of the CFGC to consider whether the gray wolf warranted listing under the California Endangered Species Act (CESA). The most comprehensive state act of its kind in the country, CESA limits takes and requires conservation recovery plans and monitoring for any animal listed under its protection. OR-7's trek had been the impetus for the Center for Biological Diversity and three other groups to petition for listing the animal under CESA in 2012. The commissioners had postponed the initial vote in February of 2014, saying they needed more time to gather information and solicit public comment.

"Reestablishment of native species is a long-term goal that excites the department. This is true for the gray wolf," underscored the California Department of Fish and Wildlife in its lengthy report prepared for the hearing that detailed the history of wolves in California. While agreeing with the inevitability of wolves repopulating the state, the CDFW recommended that the commission vote against the listing, citing their preferred alternative of a more flexible wolf management plan that didn't have the rigid restrictions of CESA, and also noting for good measure that the animal wasn't even a resident of the state at the time of

"A small, adorable conservation success story," OR-7's first litter of pups makes its debut.

Photos from August of 2015 of California's first wolf pack in ninety years.

this meeting. According to the report, "Having considered the CESA-specific factors, the Department concludes that the best scientific information available to the Department does not indicate the gray wolf's continued existence is in serious danger."

But bureaucracy had scant chance of prevailing against puppies, or toddlers dressed in wolf-ear beanies testifying in garbled words as their parents held them up to the microphone. Other children donned furry hats, and one family came dressed as a wolf pack. One individual testified in a song. More than two hundred people attended the hearing, representing a range of attitudes about wolves, from an elected official who warned that the animal is a "killing machine," to another participant who shared that "the wolf is the mother of us all." After three hours of testimony, the commission voted three to one to afford the gray wolf CESA protection, the first time an animal not living within state borders had ever been protected under the act.

Ultimately, OR-7's pups represented a promise, an aspiration, the chance for California to right a conservation wrong and position itself again at the forefront of environmentalism. As former commission president Michael Sutton remarked, his vote signified that California had a responsibility to welcome the animal home. "No species is more iconic in the American West than the Gray Wolf," he said. "We owe it to them to do everything we can to help them recolonize their historic range in our state." Californians had once purposely eliminated wolves from California; a few generations later the people of California now wanted to atone and welcome them back.

Wolf packs have always roamed throughout California, although their exact range has been the subject of much debate. "The history of wolves in California is not so obvious," writes Laura Cunningham in *A State of Change: Forgotten Landscapes of California*. Complicating the historical record is the fact that by the time we were looking for wolves, they had likely already been almost extirpated—along with much proof of their existence—the result of an aggressive and lethal predator-control policy that vanquished key parts of California's wild legacy. Yet evidence of their time here remains, and by some accounts we need look no further than a spectacular rainbow stretching over the desert.

For some of the native peoples of California, the wolf has wandered the world since the beginning of time. The Cheme-huevi people of the southeastern lands of California speak of the wolf as one of an ancient race of beings floating on a basket in the everlasting waters, accompanied by Ocean Woman, Mountain Lion, and Coyote. Together they create the world. In an epic battle in the Panamint Mountains near Death Valley, Wolf and Coyote fight their enemies the Bears. For the battle, Wolf dresses in his rainbow-colored war clothes, and the brilliant colors startle his rivals, allowing him to slay many. But a jealous Coyote betrays him, and the Bears defeat Wolf despite his valiant efforts. He eventually comes back to life, yet is banished to the north, leaving his brother Coyote regretful and howling in loneliness behind him. For the Chemehuevi, the appearance of a rainbow serves as a reminder of Wolf's discarded war clothes and his courage in battle.

For thousands of years wolves figured prominently not just in the mythology of California's native peoples but also in their daily lives. According to a study in 2013, fifteen Native American languages in California possessed distinct words for "coyote," "dog," and "wolf." The Ohlone of the San Francisco Bay Area called the wolf *maial,* while the Chumash named him *miy,* the Washoe *tulici,* and the Northern Sierra Miwok *too-le'ze.* The Pomo people of Central California still use the word *smewa,* which historically referred to the wolf, although its meaning has shifted to refer to a "hairy, ferocious, dangerous creature," according to one article on the subject.

In their poetry and myth, their language and song, Native Americans have preserved the cultural heritage of *Canis lupus* and as a result provided important information about their distribution across the state. We can add to this knowledge the clues left in accounts of early European settlers and explorers. In 1769 Spanish solider and future Las Californias lieutenant governor Pedro Fages described "deer, antelope, conies, hares without numbers, wildcats, wolves, some bears, coyotes, and squirrels of three kinds" in what is now San Diego. John C. Frémont, during an 1844 trip in the San Joaquin Valley, reported, "We saw wolves frequently during the day, prowling about after the young antelope, which cannot run very fast." William Brewer, traveling in Yosemite in 1863, recounted his party spotting a wolf near Mt. Dana, noting that it was "the only large animal of any considerable size that we have seen here." Many other records of wolves in the Golden State exist, although it's often difficult to determine in some instances if coyotes were mistakenly identified as wolves.

Based on extensive study, including a review of the historical and biological record that reinforces museum specimens and archeological evidence, the CDFW concluded in a 2012 report:

"Taken together, the available information suggests that wolves may have been widely distributed in California, particularly in the Klamath Mountains, Sierra Nevada, Modoc Plateau, and Cascade Mountains....In summary, historical anecdotal observations are most consistent with a hypothesis that wolves were not abundant but widely distributed in California."

The historical record thus demonstrates that wolves enjoyed a long tenure in the Golden State, although their exact range and population size continues to be debated. But could they survive in the California of today?

So much has changed since 1924, when the last wild wolf—an old and emaciated male missing part of his hind leg—was trapped and killed near Tule Lake, on the northern border of the state. (OR-7 missed paying a visit to this infamous place in wolf history by about ten miles.) In 1924, 3.8 million people lived in California, compared to today's population of almost 40 million. If wolves returned from myth and legend, where could they live now? What would they eat?

California is an enormous state, and although the rest of the country stereotypically envisions it filled with unending freeways and sprawl (and we certainly have more than our fair share), the congestion is balanced with a magnitude of open space. Public lands extend over almost 50 percent of the state and encompasses an area roughly the size of Florida. Within California's borders lie 26 national parks, 11 national monuments, 19 national forests, 280 state parks, and 40 national wildlife refuges. California definitely has the real estate for wolves.

But does it offer the sustenance? Carlos Carroll, a biologist with the Klamath Center for Conservation Research, says absolutely. His model, which considers factors like prey density and human disturbances, suggests that the area from the northern Sierra to the southern Cascades could support up to three hundred wolves— potentially making it the largest wolf habitat in the Pacific states.

Science says they were here. Science says they have a place to return. All these models, of course, are models, and they need testing. Perhaps the best validators are the wolves themselves, who will make clear their habitat preferences as they arrive. Even though the Sierra are not quite the wolf wonderland of the Rocky Mountains, with its seemingly endless supply of ungulates, wolves, which were once the longest-ranging land mammal in the world, certainly have the capability to survive on pretty much any landscape. Unlike coyotes, they avoid urban areas, and also prove extremely good at self-regulation—pack size is moderated by prey availability. Given the shrinkage of their traditional range, and the reduction of elk herds, it's unlikely California will ever support a large wolf population, and yet since wolves inhabit just 5 percent of their historical habitat across the United States, every little bit helps. And most Californians seem eager to be part of the effort; polls show 80 percent of residents support wolf-recovery efforts.

I like to think that the next generation of Californians will see wolves in rainbows. That they'll share a living memory of wolves, that they'll invent new stories and slangs and myths. Three tribes in Northern California—the Yurok, Karuk, and Hupa—to this day perform a series of ancient religious dances designed to "fix" the world, to restore balance and foster renewal. As part of the White Deerskin Dance, participants wear wolf-skin blinders over their eyes to signify the wolf's role in the "balancing of the world for all things." Will the new Californians create a dance to honor the wolf? Will the wolf be a trickster or a benign creator? The particulars of the wolf's rebirth in our cultural mythos doesn't matter. What matters is for the wolf to be here to inspire it.

"I will try to create a feeling for wolves that we may once have had as a people ourselves but have long since lost—one in which we do not know all the answers, but are not anxious."—Barry Lopez, *Of Wolves and Men*

"As California is the largest state in population, and has a more diverse and urban population than any other wolf state, the Department believes the tolerance for wolves overall in the state would be high (this is supported by the level of comments received from the public on this topic)."
—California Department of Fish and Wildlife report (2014)

A full moon dominated the night as I drove through Yellowstone National Park one November. The white landscape amplified the moonlight, and as winter moved in, snowdrifts and ice were beginning to take their rightful place, including on the road. In another couple of weeks, bison, elk, and other animals would reign over Yellowstone, as the park would be closed to automobiles almost entirely as it shifted to winter use.

In the back seat slept my dog, Sasha, old and slowly dying from cancer. The storm had passed, the last snowflakes falling gently. Ahead, I saw movement in the shadows—something emerging slowly from the forest. I slowed my timid speed to a stop as three large figures stalked toward me. Too small for elk or bison. In the glow of my headlights, I saw the unmistakable faces of wolves.

I caught my breath and quickly pulled out my camera, always handy in the front seat, a lesson you learn when living in Yellowstone. Surprisingly, they headed directly toward me. They stopped in the headlights for a moment, then turned to walk around the driver's side of my car.

I could barely breathe. As Craig Childs describes in his *Animal Dialogues,* "Times that I have seen the animals have been like knife cuts in fabric. Through these stabs I could see a second world." The night was so quiet I could hear the pads of their feet crackle in the snow. Snowflakes clung to their fur, two with black coats and one gray. My friend and longtime wolf campaigner Amaroq Weiss put it this way: "They are big animals, and I don't mean physically—they have an enormous presence and you feel them."

My twelve-year-old shepherd mix, raised in environs totally devoid of *Canis lupus,* woke up suddenly. Although she had become lame and standing was painful, she stood upright in the confident stance of youth, ears fully alert, sniffing through the crack of the open window as they passed. She didn't bark or whine like she would have if she had seen another dog. Her gaze seemed admiring.

I rolled down the window about a quarter of the way and snapped some photos as one of the black wolves passed, then another. Too dark for my camera. Then the gray approached, and she glowed in the moonlight. She was generous to me—she looked up and paused, if not curious at least acknowledging my presence—and allowed me to capture her on film before she moved on. Immersed in rapture, I felt like the poet Kotomichi, who said, "My heart was rapt away by the wild cherry blossoms— will it return to my body when they scatter?" I watched them disappear into the trees.

Wolves excite our passions in a manner that no other animal does. They inspire love as much as hatred, reverence as much as fear. As such, for wolf restoration to work, we need to develop a shared social compact about wolves. Science tells us what is technically possible—the biological conditions exist for wolves to once again inhabit California—yet there is a social component to our relationship with wolves (and all other wildlife) that cannot be ignored. And let's be honest, many worthy and unique species went extinct because the political or public will wasn't there to save them, despite reams of scientific papers. As a society we have to agree that wolves, or California condors, or red-legged frogs, are worth protecting if they are going to endure. As we've observed in the ongoing wolf wars between those who want

The author's memorable encounter with a wolf in Yellowstone National Park.

wolves on the landscape and those who do not, scientific facts
have not budged people much from their various factions.

My intent isn't to pass judgment on either side. I want to
explore what is possible. As John Shivik asks in his recent book,
*The Predator Paradox: Ending the War with Wolves, Bears, Cougars, and
Coyotes,* "The rapid sprawl of civilization forces the issue: Is there
anywhere else for predators to go if they can't live on humanity's
doorstep? Are there options that would allow us to have carnivores in our kingdom while we protect our livestock, property,
and people? Finally, who is going to jump in the fray between
people and predators and end the feud?" At the California Wolf
Center (see page 169), the director of California wolf recovery,
Karin Vardaman, sees reason to hope. She's part of a broad task
force called the California Wolf Stakeholders Working Group,
which includes diverse perspectives from ranchers to environmentalists, as well as a smaller collaborative effort to develop the
California Wolf Livestock Coexistence Plan. Both groups work
on developing proactive solutions to minimize conflict as wolves
return to California.

"What I have seen makes me very positive for the future of wolves in California. We have been working toward building positive relationships and trust that will both protect livestock and save the lives of wolves in the wild," she says. "Not all are excited about wolves but understand they are a reality. Our common goal is that we don't want to have livestock die because of wolves and we don't want wolves to die because of livestock. Many ranchers here are open to nonlethal solutions, and what is going to be important is providing them with resources for implementing these solutions."

One of those ranchers is Keli Hendricks. She learned about the land from her father and uncles, who were raised on a farm in Idaho. She now lives with her husband in Sonoma County, where together they oversee a cattle operation. They share the land with many predators, such as coyotes, bobcats, and mountain lions, but as Keli jokes, "The fiercest animal I have dealt with, and the only one to actively attack, has been a western gray squirrel. People think I am kidding, but I'm not."

Keli trained cow horses for many years and worked on a number of ranches before making her home in Sonoma County. She witnessed a variety of different attitudes toward wildlife in her travels, and joined the board of Project Coyote to help educate the ranching industry about how to coexist with predators. "I love our coyotes," she says. They keep our ground squirrel populations in check and they are just overall great neighbors. Working this closely with predators has made me appreciate them even more. Most people have no idea how beneficial and nonaggressive these animals are. Even the smaller predators like skunks, opossums, and raccoons eat tons of insects, rats, and mice."

And for her, wolves are no exception: "We would welcome wolves. We don't look at wolves as a nemesis, or as friends, we look at them as just another species with which we would learn to share the land."

Filmmaker Daniel Byers is also seeing evidence of California and the Pacific West becoming models for a new wolf country. He assembled a group that included a wolf ecologist, a National Geographic Young Explorer, and a wild peace advocate to retrace twelve hundred miles of OR-7's footsteps for the documentary film *Wolf OR-7 Expedition*. "For me—having seen a lot of polarizing discussions around wildlife conflict issues worldwide—I was particularly attracted to the interesting middle grounds in this story, the potential to break old narratives and explore new ones," he says. During their hike, they visited with people in this new wolf frontier—cattlemen, hunters, ranchers (one a combination rancher and three-time Olympic cyclist)—and considered the question What does it mean to live with wolves? And Daniel discovered that the old stereotypes didn't apply in most cases: "In our journey we consistently encountered people who broke the traditional model of wolf hater or wolf lover, people who embraced the idea of science-backed coexistence."

For wisdom in all things wolves, I always turn to my friend and mentor Douglas Smith, who has headed up the Yellowstone Wolf Project since 1995 and has authored several books on the animals. When OR-7 finally took his first footsteps into California, we celebrated together via email. I asked him about the likely success of wolves repopulating the Golden State. "California is a great example of the adventurous spirit that wolves have," he said, "and it also shows that wolves don't need much; give them a little bit of a break and they'll do the rest."

In August of 2015, the wolves themselves ended all the speculation and debate. The California Department of Fish and Wildlife announced the discovery of two adults and five pups that had formed a pack in Northern California; as before, they defied the expectations of a human timetable. "We were really excited, if not amazed, at the appearance of the wolves. They have beat us to the punch," exclaimed Eric Loft, chief of the Wildlife Branch for the California Department of Fish and Wildlife, to the *Los Angeles Times*. Named the Shasta Pack, this group is even more remarkable because it did not originate from OR-7's nearby Rogue Pack but rather from his original family in northeastern Oregon, perhaps even following the scented trail OR-7 left on his journey. However they arrived, though, the word is out in the wolf world, and after ninety years, California is now an official wolf state once again.

"California is ready for wolves," says Amaroq and I agree. And ready or not, they have arrived. I don't know what this new wolf country will look like, but I do know California will figure it out. We've figured out how to coexist with earthquakes—wolves should be a breeze. We've proven to be good neighbors to a lonely mountain lion in Los Angeles and want to build the largest wildlife crossing in the world in his honor. We sparked the modern environmental movement while restoring an entire waterway for porpoises and other wildlife in San Francisco, we have begun reenvisioning the built landscape of Silicon Valley into a Nature 2.0 that's a good enough compromise for foxes, egrets, and other creatures, and we have taught bears to be wild again in one of the largest national parks in the country, simply by correcting our own behavior. Pioneering a new "wolf country" could be our next contribution to conservation—and the lasting legacy of OR-7's pilgrimage back to the land of his ancestors.

The California Wolf Center

The town of Julian, California, is best known today for two things: the California Wolf Center and the Julian Pie Company's scrumptious pies. On my first visit, I sampled both. Even before OR-7 entered the state and inspired wolf-recovery efforts, California contributed significantly to wolf conservation across the United States, an effort that can be traced back to this small mountain community fifty miles east of San Diego.

Formed in 1977, the California Wolf Center started out as a general wildlife education center, but it shifted its focus to working on the reintroduction of the Mexican gray wolf, one of the rarest land mammals in the world. The center breeds this subspecies of wolf for release in the wild. Erin Hunt, the center's director of operations, gave me a tour of the facility and introduced me to their ambassador Mexican gray wolves. "These wolves will never be released but play such a critical role in education," she says. "People who tour the facility get so inspired by seeing these animals." The wolves intended for rewilding were kept out of site, but I could hear them howl at times.

Today, the center continues its vital efforts with the Mexican gray but has also added a major emphasis on California wolf conservation, inspired by OR-7's historic trek. "He was a game changer," says Karin Vardaman, who leads the center's California wolf-recovery efforts. Instead of breeding wolves for release, this program focuses on developing proactive solutions to make the Golden State a safe haven when wolves arrive. She's excited about the project, and clearly loves her work with wolves as well: "I remember my first days on the job here—it sounds silly, but I got excited about picking up wolf poop."

A wolf and pups at the California Wolf Center.

OR-7: The Movie

"The audience will fall in love with OR-7!"
"The epic story of a lone wolf's search for love."
"This film promises to touch the heart."
Filmmaker Clemens Schenk might not have known much about wolves before making this film, but in OR-7 he knew a star was born. He watched the headlines about the lone wolf as the animal neared California, and one weekend he read through all of the news articles he could find. "I was so surprised at the immense terrain this wolf crossed, and the amount of strength and stamina he needed," he remembers. "It fascinated me that one animal could capture worldwide attention. He's a rock star." Clemens became convinced the story had all the makings of a blockbuster film and that OR-7 made a natural hero.

His instincts were correct, and when the movie was released after two and a half years of production—which required strenuous hiking through some of the remote areas where OR-7 traveled—it screened to sold-out theaters and became an official selection of the Wild and Scenic Film Festival. Distribution has steadily increased, with scheduled showings on PBS and thousands of viewers watching it online. For Clemens, the attention the film is receiving is gratifying, not because he desires accolades as a filmmaker but because he hopes it will help educate people about the plight of wolves and "clear up misconceptions and myths about these magnificent animals." He considers OR-7 the perfect ambassador to the general public.

Since a wild wolf like OR-7 couldn't be expected to cooperate with a film schedule, who did Clemens cast in the canine role of a lifetime to play the celebrity? After auditioning several captive wolves, he found his star at the Wolf People Reserve: a playful animal named One. "As soon as I saw him, I knew he was my wolf. He looked like OR-7 and was the clown of the pack."

The movie poster for *OR-7: The Journey.*

Buddy the Wolverine: Another Lonely Wanderer

"Who runs into wolverines?" asks Douglas H. Chadwick, author of the riveting book *The Wolverine Way*. A rare few may encounter what he calls "the toughest animals in the world," and it's probably for the better. This thirty-pound animal is notoriously fierce and, despite its diminutive size, has been known to chase off grizzly bears from their kills and hunt much larger animals, such as moose and caribou. Opportunistic feeders who dine on both live prey and carrion, the wolverine fits its scientific name, *Gulo gulo,* which translates into "gluttonous glutton."

Wolverines had not been spotted in California since 1922, and not many people held out hope for their return. Then in 2008, a remote camera captured one wandering in the Sierra Nevada just north of Truckee. Sierra Pacific Industries biologist Amanda Shufelberger nicknamed the wolverine "Buddy," and she and another biologist with the California Department of Fish and Wildlife, Chris Stermer, have now amassed more than a thousand photos of him to date. He appears healthy and thriving, but he's likely suffering from lovesickness, as no female has been spotted in the vicinity.

Buddy—and all wolverines—are solitary and prolific wanderers. He likely traveled five hundred miles from Idaho to reach his new home, and the territory where he now roves spans about three hundred miles. Since a long-distance relationship doesn't appear to be a problem for a wolverine, though, a mate may arrive and surprise both him and us, but time is running out as he approaches the average lifespan for his kind. For now it's enough to celebrate his presence; as Chris Clarke, an environmental writer for KCET, wrote, "It's nice to know that at least one big, very wild animal is finding enough room in the Sierra to go about its business without us bothering it."

"Buddy" is the first wolverine seen in California since 1922.

Bringing Back Salmon with the California Conservation Corps

When I spoke to John Griffith, longtime crew supervisor for the California Conservation Corps (CCC), he had just returned from five months in the backcountry with his team working on fire response and salmon habitat restoration. "Salmon is by far my favorite project," he says. "You actually get to see what you are saving while you are saving it. While the CCC members are anchoring logs in streams, they see salmon swimming around their feet."

For more than a decade he's supervised hundreds of college-aged young people recruited to perform natural-resource work as part of this government program that advertises "hard work, low pay, miserable conditions." For many crewmembers, this is the first time they have even experienced the natural world—many of them have seldom played outside prior to their tour. "One student had spent every free moment in front of the television or computer before joining the CCC," John said. "He actually thought salmon was something that came out of a can. But then he became interested. He got a snorkeling mask so he could see the fish, and soon got promoted to Salmon Habitat Specialist. Part of the reason I love my job is we're creating future wildlife heroes."

Salmon will need all the champions John can recruit. If California's changing climate persists, researchers believe coho and Chinook salmon will be extinct in one hundred years. In 2015 the state had to truck thirty million fish to their spawning grounds, as the rivers lacked enough water for the fish to swim. But despite the challenges, John remains optimistic: "I know restoration works." He tells of a "dead" stream site he rebuilt after it had experienced devastating clearcutting and floods. A decade later, he snorkels there with salmon, turtles, and river otters—evidence that if you build it, wildlife does indeed come.

John Griffith leading a CCC crew in salmon restoration.

"Refrogging" California by Building Wetlands

Mark Twain featured California's red-legged frog in his famed 1867 story "The Celebrated Jumping Frog of Calaveras County," the inspiration for today's Jumping Frog Jubilee held annually in Angels Camp. But don't expect to see the red-legged frog in the contest—instead, bullfrogs, nonnative to California, leap across the stage for the top prize.

Why the absence? Now gone from 70 percent of its historical range, the once common red-legged frog owes its decline to miners who used them for food during the gold rush. The challenges today—ranging from invasive species (including the bullfrog that preys upon it) to disease, drought, and habitat degradation—have driven the species to near extinction. Habitat loss is especially critical, as development has destroyed 90 percent of the state's wetlands, which are vital breeding grounds for the frog.

Kerry Kriger, founder of Save the Frogs!, is on a mission to "refrog" California and restore the native creature to its once abundant numbers. He counts the National Wildlife Federation as one partner in this effort. Kerry, who also led a successful campaign with the students of Sea View Elementary School to designate the red-legged frog as the official state amphibian in 2014, views his wetland restoration workshops as an easy solution with long-lasting benefits. "Constructing wetlands is a fantastic method of educating students, teachers, and community members about amphibians and ensuring that amphibians have a home in which to live and breed," he says. His expert design partner, Tom Biebighauser, agrees: "Wetlands can last hundreds or thousands of years. There are very few things we build in life that last that long."

One of their first projects in Shingle Springs reclaimed a wetland that had long been drained for agricultural uses. The landowner noticed an immediate transformation. "The pond is doing well," he says. "Lots of tadpoles and birds visit. We love it."

California's native red-legged frog.

Ranchers Helping Ringtails in the Sutter Buttes

Professor David Wyatt, who once majored in criminal justice but switched after taking a natural history class, now teaches in the field ecology program at Sacramento City College. Since 1985 he's been studying ringtails—members of the raccoon family—in the Sutter Buttes, deemed the world's smallest mountain range.

What drew you to study ringtails? As crazy as it sounds, getting bitten by a wild ringtail gave me the bug. I had just transferred to Sacramento State University from a community college and took a class with the man who would become my mentor, Dr. Gene Trapp. One day he asked, "Does anyone want to go out and live-trap ringtails?" *What in the heck is a ringtail?* I thought and, being curious, I volunteered, and on that first trip got bitten by one by accident. I love this work and I especially love the glimpse we have into the lives of this amazing and charismatic mammal. Very little is known about them, and that's what also draws me.

What is your research showing you? We're finding that ringtails in the Buttes are mainly vegetation eaters—even though the species are omni-vores—based mainly on food availability. And that they have tiny home ranges in this area.

You do some of your most important research on a cattle ranch. How did that come about? While we were doing other field work in the Sutter Buttes, we met Margit and Pete Sands, who own a ranch that has been operating since 1898. After they learned of our study, they told us about the ringtails on their property and gave us free access for our research. They are very active stewards of their land, take really good care of it with a low-impact oper-ation, and like sharing their home with wildlife. Without Margit and Pete, none of this ringtail work would have been done in the Sutter Buttes, and I fully credit them with being the catalyst behind this research.

A ringtail in the Sutter Buttes.

Rice Farmers Sharing Their Fields with Sandhill Cranes

Nature choreographed an elegant ballet in the dance of the sandhill crane; the graceful birds flap their wings, leap and hop into the air, and bow their red-crowned heads. Although primarily a mating ritual, it's also performed outside of courtship. The first time I witnessed this energetic dance was during the Sandhill Crane Festival in Lodi, an annual event celebrating the cranes' return to their winter home.

The sandhill crane covers twenty-five hundred miles in its annual migration, and California's Central Valley is along one of the main routes. Plentiful flocks of sandhill cranes used to darken the skies in the Golden State, but by the 1940s the breeding population had been reduced to fewer than five pairs. Today, their population has partially rebounded, but it is nowhere near what it once was.

One reason for their decline involves the widespread disappearance of wetlands, largely to development and agriculture. Some agricultural interests, however, are now helping restore lost habitat. Through the Migratory Bird Conservation Partnership, organizations including Audubon California, the Nature Conservancy, and Point Blue Conservation Science partner with California farmers who grow rice and other crops to manage their lands in bird-friendly ways. Audubon California's executive director, Brigid McCormack notes, "Right now, rice represents nearly 80 percent of flooded habitat used by migratory birds in the Sacramento Valley."

Working with the National Resources Conservation Service and the California Rice Commission, the partnership launched a pilot program in 2011, and today 20 percent of area rice farmers are on board. In a report on the project, Keith Davis, one of the rice growers participating, commented, "I am really happy with this new habitat program....I've seen increased numbers and variety of species using my fields. I love driving around and seeing all these birds."

Sandhill cranes gather on the San Luis National Wildlife Refuge Complex.

Coyotes on the streets of Los Angeles.

Good Neighbors

WHAT CALIFORNIANS ARE DOING FOR WILDLIFE IN THEIR OWN BACKYARDS

Imagine neighborhoods uniting to create a milkweed path across an entire city to help save the monarch butterfly. Or a frog pond in every schoolyard—a living biology class that also increases California's native amphibian population. Or apartment buildings with edible hanging gardens that feed both humans and hummingbirds.

When people ask me what they can do to personally help wildlife, I always answer, "Make your home or business wildlife friendly." One of the easiest ways we can all ensure a future for our fellow creatures is by creating numerous habitat areas in our backyards, schoolyards, corporate properties, community gardens, parklands, and other spaces.

Since 1973, the National Wildlife Federation has provided millions of people with the basic guidelines for making their landscapes more wildlife friendly through its Garden for Wildlife program. By planting or preserving the native plants that wildlife need to survive on our properties, and providing natural sources of food, water, cover, and places to raise young, anyone can create a special kind of "garden" for local wildlife. The NWF will recognize such landscapes by designating them as Certified Wildlife Habitats. When entire cities commit, they can be certified as Community Wildlife Habitats. Whether you have an apartment balcony or a twenty-acre farm, you can create a garden that attracts wildlife and restores habitat in commercial and residential areas.

NWF naturalist David Mizejewski is a champion of the Garden for Wildlife program because everyone can participate. "It is easy to feel as if there is no hope for wildlife in our modern world of asphalt, smog, and traffic. But there is hope," he says. "You can choose to create a garden or landscape that helps restore the ecological balance in your yard. You can surround yourself with beautiful native plants that will attract wildlife and allow you to observe an amazing array of wildlife every day."

I offer as inspiration these stories of people across California who have inspired me by welcoming wildlife to their neighborhoods. Join them, and the community of more than 190,000 people nationwide, who have committed by enrolling their landscapes and gardens as Certified Wildlife Habitats. By transforming our backyards, schoolyards, and communities into wildlife-friendly spaces, we help both people and animals thrive.

CRAIG NEWMARK

Like many other nonprofit and social media enthusiasts, I started following Craig Newmark on Facebook for his technological genius, witty commentary, and dedication to philanthropy. Yet in between his eclectic posts, his dedicated advocacy for veterans' rights, and even answering customer questions for Craigslist (his official title reads Customer Service Representative and Founder), I started noticing a curious pattern.

Squirrels.

Photos of squirrels gathering at a birdbath or raiding a birdfeeder. Or the *Mission: Impossible*–esque squirrel surveillance and reconnaissance video series, in which Craig applies his high-tech skills to catch some undeniably cute home intruders red-handed. But this interest in squirrels is not cursory; in 2012 he initiated a #Squirrels4Good fundraising campaign for the National Wildlife Federation, as part of his larger effort to create "a new era of squirrel-based activism."

Why the affection for squirrels? As Craig told me in an interview, "Squirrels are survivors; they impressively adapt to the urban environment. My favorite encounters are caught on webcam, where I can see one very smart squirrel coming in[side my house]

One of Craig Newmark's many backyard squirrels.

to check things out. No interior video, but, well, after seeing the video, I see something to the right of my keyboard, and let's say... well, that's not a raisin."

Craig's "mi casa es su casa" philosophy with the local wildlife means it's not just squirrels that might wander into his home and leave some mementos. Craig also regularly shares his photos capturing an array of other backyard critters. "I live in the Cole Valley neighborhood of San Francisco, in an odd side street that backs into Sutro Forest. Sutro Forest is an actual forest in the middle of the city, near Golden Gate Park. It's large enough to have surprises, like a coyote sighting by two different people. It has hiking trails; I've been up there twice with the missus."

Craig's wife, Eileen, shares his bird- and squirrel-watching enthusiasm, and he gives her full credit on Facebook when she spots a life-lister bird or snaps a photo. Craig's roster of avian sightings in his backyard alone is impressive—fifty-three species and counting as of June 2015—and he documents them in an online photo collection titled "Eileen and Craig's Birdography Spectacular." If the technology thing ever fizzles, Craig could easily switch to wildlife photography as a profession.

Given the abundance of critters he sees, what are some of his most memorable encounters? "For me, maybe the first times I saw the most unexpected of birds, particularly hummingbirds and various raptors. For the missus, the times when a tree rat darted out from behind a plant pot, or maybe more recently when a pair of raccoons tried to break in."

What animals would he like to add to his wish list of backyard visitors? "I'd like to get a good sighting of the coyote; we could hang out and be pals, though Crosby, the next-door terrier, wouldn't be so keen about that."

I'm guessing the squirrels might not like it either.

ELIZABETH SARMIENTO

Elizabeth Sarmiento, an avid urban gardener and environmental activist, remembers being struck by the typical monoculture landscaping in the United States: "Back in Honduras, where I am from, gardens are not manicured things like square lawns and flowerbeds. Gardens are ecosystems with an abundance of different plants that feed people and wildlife."

After moving to San Jose to join her fiancé, she started transforming his urban home into a chain of microhabitats that support threatened birds, amphibians, butterflies, and pollinators. She removed the front and back lawns and irrigation systems and replaced it with a wildlife habitat garden with sages and other plants, as well as adding stones and logs to make homes for

Elizabeth Sarmiento, cofounder of Smart Yards Co-op, pruning in her urban eco-space; her niece and nephew enjoy plums from her garden.

wildlife. "I was so excited when the first lizard showed up scurrying up one of the logs just a few months after building the garden," she said.

Her small backyard is dedicated to growing food—and an abundance of it—such as pears, apples, oranges, kumquats, figs, plums, mandarins, chilis, cauliflower, lettuce, arugula, endive, romaine lettuce, peas, and chamomile. She also designed an area for composting and built a chicken house. But wildlife also frequents this space. "My backyard is a sanctuary, where I sit still every morning with my cup of coffee to observe and to enjoy nature with gratitude. Birds, butterflies, bees, and amphibians have food, shelter, water, and all they need to thrive year-round in the garden," she says.

Elizabeth became so passionate about the project that she took a workshop on native landscaping and now trains others to convert their yards into urban gardens that feed both people and wildlife. Fostering thriving communities is a focus of her work as a Latino community organizer, and she considers urban gardens a way to create healthy spaces for people. She works with a variety of organizations and agencies on promoting sustainable practices and tries to infuse the importance of creating gardens and green-space into all of her advocacy. As director and cofounder of Smart Yards Co-op, the first sustainable landscape cooperative of its kind in San Jose, Elizabeth gets to design, teach about, and, with a team of workers, install California native habitats, graywater systems, and drought-tolerant gardens that foster biodiversity and conservation while also meeting our human needs all over the San Francisco Bay Area.

"To me, urban gardens are such an easy solution, and I try to show people by being an example of what is possible in a city," she says. "Teaching people about things like composting and how to plant diverse native gardens creates healthy neighborhoods. It results in a lot of food for people—and for wildlife— year-round."

THE TOWN OF ALPINE

Residents of the small hamlet of Alpine applied at such an unprecedented rate to have their yards recognized as Certified Wildlife Habitats that they inspired the National Wildlife Federation to create a new award level for its participants. On May 1, 1998, Alpine became the country's first NWF Community Wildlife Habitat. Today, more than eighty-three communities across the United States have completed the two-year process to achieve the national status, including Austin, Texas; Bethlehem, Pennsylvania; and Annapolis, Maryland. Dozens more are currently working toward certification.

Why the deluge of certifications in Alpine? Avid gardener Maureen Austin hosted a local party in her certified backyard that provided the inspiration. "It was in the late afternoon, near dusk," she recalled in an article for *National Wildlife.* "Hummingbirds, songbirds, and butterflies were everywhere. The wildlife really intrigued everyone. They found it enchanting." People soon followed her example, and certifications increased significantly, quickly earning the city the record of the most certified habitats in the country at the time.

Maureen served for nineteen years as the executive director of CHIRP (the Center to Help Instill Respect and Preservation for Garden Wildlife, Inc.), whose mission was to educate people about gardening for wildlife. As Maureen notes, Alpine's Community Wildlife Habitat award from the NWF has been very beneficial to local wildlife but also to community spirit, as it has given residents "a positive identity—something to be proud of."

Alpine was the first NWF Community Wildlife Habitat in the country.

SUSAN GOTTLIEB

The garden is a love song, a duet between a human being and Mother Nature," wrote *Landscaping with Nature* author Jeff Cox. A perfect example of this sentiment is the Gottlieb Native Garden, whose creator, Susan Gottlieb, has proven herself a talented collaborator with the natural world. Their ongoing duet has resulted in one of the largest private native gardens in the country—one that has been featured in the *New York Times* and the *Los Angeles Times,* and is the subject of an upcoming book.

Entering her garden, you discover an unexpected wonderland. From her property in Beverly Hills, Susan possesses a panoramic view of downtown Los Angeles, complete with skyscrapers, trafficked roadways, and sometimes the haze of the city's infamous smog. Yet as you meander on her terraced path, passing madrone trees and desert lavender blooms, you have a hard time believing she is surrounded by the second-largest city in the country. The music of a small waterfall lulls you immediately into a state of relaxation, a perfect backdrop as you pass through a colorful cloud of hummingbirds that buzz and whirr as they pass her collection of feeders. Descending down the hill, beyond the bluebird house boxes, you hear the steady hum of native bees and can sense their delight as they feast on the rainbow of snapdragons, sunflowers, and poppies.

Once, a tangle of nonnative vines and tropical plants covered this hillside property, a typical Los Angeles garden that had been designed for decorative effect. That changed when Susan became engaged to the owner, Dan Gottlieb. Dan jokes about his wife's passion, "Most girlfriends tear up your black book. She tore up my yard." The impetus for the transformation? "I wanted to see more birds," remembers Susan. "Growing up in Ontario, my mother used to feed the birds, and I developed a lifelong passion for bird watching."

A list of avian sightings in her backyard would be the envy of any birder: she's seen a great horned owl, a great blue heron, a ruby-crowned kinglet, and California quail, to name a few. A diverse array of other wildlife have also visited her sanctuary, including deer, skunks, woodrats, bobcats, gopher snakes, cotton-tail rabbits, fence lizards, bats, and coyotes. This abundance of city fauna has also attracted the notice of scientists, and her garden has been the site of several research projects, including a recent study of urban bats.

On her one-acre urban hillside property, Susan has clearly demonstrated that a little goes a long way in attracting wildlife. She's also demonstrated that a little goes a long way in terms of

Avid gardener Susan Gottlieb; a female phainopepla takes advantage of one of her birdbaths.

water usage—another factor that inspired her to learn more about native plants and sustainable gardening. "I'm a nurse by profession, and I knew very little about landscaping. When it was clear that the conventional gardens so prevalent here use copious amounts of water, I started to investigate," she says. "I read books, reached out to experts, and used resources like the Theodore Payne Native Foundation for Wildflowers and Native Plants."

Susan's passion for wildlife and wildlife-friendly gardening and her love of photography—interests she shares with her husband— led them to open the G2 Gallery in Venice, California, in 2008. Their award-winning gallery of nature and wildlife photography showcases photographers from around the world and donates proceeds to environmental charities. Active in many environmental causes, she is determined to extend her efforts beyond her backyard sanctuary. "The gallery and the garden—I want both to inspire others to take action to help wildlife," she says.

CALIFORNIA CONSERVATION CORPS CAMPUSES

Founded in 1976, the California Conservation Corps is the oldest and largest conservation corps now in operation. Since that time, 120,000 young people—hailing from around the state and reflecting the diversity of California—have signed up for a job that advertises "hard work, low pay, miserable conditions...and more!" Together they have performed sixty-nine million hours of natural-resource work, from restoring salmon habitat in the backcountry (see page 172) to planting trees in urban parks.

John Griffith, who is a veteran of the program himself and has led crews for thirteen years, loves the program's ability to forge connections between nature and young people. "A lot of times these crew members have never seen wildlife before," he says. "When people join CCC, one big thing is they don't want to get their feet wet. By the end of summer, the same people are putting on a snorkel and mask and reporting every salamander or fish they see. I like my work because through my crews I am helping create twenty-five wildlife heroes a year."

In 2013, one of John's crews took that dedication to wildlife a step further. Under his direction, they worked to create wildlife habitat on their campus in Ukiah. John, who started reading *Ranger Rick* magazines in first grade, had invited me to spend some time with his crew to talk about wildlife conservation. When they learned of the NWF's Garden for Wildlife program and its Certified Wildlife Habitat designation, they immediately knew this was a way to use the restoration skills they had learned in the backcountry to transform areas in their communities into wildlife-friendly spaces. Corps members and staff in Ukiah planted berry-producing plants, trees, and bushes, and center director Mark Hill secured the donation of a fountain. They

CCC leader John Griffith (right) and crew member Casey Esparza.

installed bird and bat houses, and created pollinator pitstops for migrating hummingbirds and resident butterflies.

Crew member Kevin Casbeer, who helped with the campus restoration, loved the project: "We got the rad opportunity to help create and restore habitat for wildlife—not only rural areas but in residential settings as well. It was a great way for us to have a fun and cool connection with nature, which is what the CCC is all about!"

Since then, the program has expanded, and in 2015 the Fortuna campus announced its certification—its habitat now complete with a birdbath made out of a crew safety helmet and broken tools. John plans to help all thirteen CCC campuses certify and to expand opportunities for corps members to learn about the NWF's gardening-for-wildife message.

SAN FRANCISCO AIRPORT MARRIOTT WATERFRONT

In 2012, a guest at the San Francisco Airport Marriott Waterfront reported a rather unusual visitor to hotel management. People staying at the hotel expect views of the San Francisco Bay from their rooms, but they didn't anticipate opening their window curtains and seeing two barn owl parents feeding their endearing baby owls on the other side of the glass.

Many businesses might have treated the owls as pests (it is illegal to remove active nesting birds, but adding barriers to prevent nesting is common), but the Marriott staff chose instead to roll out the welcome mat to the family. Clif Clark, the general manager of

Barn owlets check in to the "Hoot Suite" at the San Francisco Airport Marriott Waterfront hotel.

the property, said they never even considered disturbing the owls: "The hotel is by the bay, and we embrace the beautiful nature at our location."

The owl parents made the small ledge outside the window of room 1141, nicknamed the "Hoot Suite," a nursery for four owlets. And the hotel employees embraced their role as nursery caretakers. They educated guests about their avian roommates, teaching them how to be kind to these feathered friends (don't open the window or flash light at the owls), and they offered a gift pack for children who stayed in the Hoot Suite that included a stuffed toy owl. "I love showing our younger guests the baby owls and giving them a toy owl to keep as a memory of the event," says Clif.

After seeing the photos of the baby owls on social media, I asked the hotel for a tour. The director of operations, Dean Waziry, accompanied me and patiently stood by as I oohed and aahed over the impossible-to-resist fluffy baby owls. "I never get tired of looking at them either," he commented. "We love having them at our hotel." Even though I see wildlife on a regular basis, I'll admit I could have watched the captivating owls for hours. Two of them napped while the other two gazed at me through the window with hopeful expressions, perhaps wondering if I had some rodents to feed them.

The owlets' room service would not arrive until later that evening, as barn owls are entirely nocturnal and stalk their prey in the night world. Although they have excellent vision in the dark, David Lukas, author of *Wild Birds of California,* notes that barn owls possess a "refined ability to locate prey by sound." As the most widespread of all owls in the world, barns owls are easily identified by their ghostly white color, heart-shaped faces, and raspy, screeching calls.

Although they are the most common owl, the Cornell Lab of Ornithology notes that their numbers "seem to be in decline over much of North America (and Europe)." As per their namesake, barn owls like to frequent barns and other human-made structures, but you don't need a barn or a hotel building to create a home for these owls, who also provide excellent natural rodent control. Build it and they will come! Place a nest box on a tree or a pole in your yard and you might soon have a nesting family of your own.

As barn owls often nest in the same place each year, it's likely these owls may now consider the Marriott a regular destination, although in 2015 they seemed to have skipped a year. For the staff and guests, the owls have become a regular part of the hotel experience when they are nesting. As Clif observes, "Many times I have seen the owl parents around dusk. It is great viewing while sitting at our outdoor fire pits that overlook the bay."

As hotel amenities go, a live viewing of cute baby owls certainly outranks free HBO or Wi-Fi, although TripAdvisor does not yet have a designation for owl watching. Forget spas; perhaps wildlife viewing can be the next must-have offering for a hotel? This is fine with Clif. "We have plenty of space for more owl families," he says. And what other critters would he like to welcome next? "Maybe an eagle family would be nice."

LEO POLITI ELEMENTARY SCHOOL

Leo Politi Elementary School is located in a densely populated area of Los Angeles and surrounded by concrete, roadways, and buildings. Yet birds such as the Cooper's hawk soar over its playground, and the song of a northern mockingbird might be heard through a classroom window.

Nature is an essential part of the curriculum at Leo Politi Elementary, and it has been since 2008, when, under the direction of its principal, Bradley Rumble, it transformed a concrete schoolyard into a wildlife-friendly habitat through a project led by Los Angeles Audubon with a grant from the US Fish and Wildlife Service. Classrooms are stocked with *The Sibley Guide to Birds,* and students at recess can identify the resident Anna's hummingbirds.

Students in the wildlife garden at Leo Politi Elementary School.

Margot Griswold, now the president of LA Audubon, was a scientist with the organization during installation of the native habitat, and she has seen firsthand how the appearance of wildlife has engaged the students. "One of the best wildlife sightings was an Argiope orb weaving spider that had a perfect web in a California sage shrub at the edge of the habitat where many students could observe the spider feeding," she remembers. "That day we had many impromptu lessons as students would bring their friends to see this amazing, large, yellow-and-black female spider. All the students were very respectful of the spider."

Installing the wildlife habitat also paid dividends inside the classroom: test scores rose significantly after its completion. But dedication to nature doesn't just end when the school bell rings and class is dismissed. Giving students a connection to nature has far-reaching impacts, as Rumble notes. "The students, staff, and families are learning about the interconnectedness among all aspects of an ecosystem," he says. "As we do this, we also learn about the interconnectedness of all of us, and that's an important life lesson in this enormous city."

Rumble, who has now moved to Esperanza Elementary School, shares a vision for creating wildlife-friendly spaces in all Los Angeles schools. As he wondered aloud to *SoCal Wild,* "Why couldn't all the schools in Los Angeles become an urban flyway for migrating birds? Every school in every neighborhood. That would really be remarkable. We could all be connected with the birds."

PETER COYOTE

Yes, his namesake animal does wander into Peter Coyote's back-yard, along with a diverse array of critters including skunks, gray foxes, raccoons, and birds galore.

His wild menagerie attests to the actor's lifelong affinity for nature. "I've always been fascinated by animals and have felt a kinship," he says. "By the time I was eight years old I realized that everything in the world was alive and connected, and had its own business—and you didn't interrupt it without consequences."

A resident of Northern California's Marin County since the 1970s, Coyote has witnessed some of the negative consequences of our actions on the natural world and considers his efforts for

A gray fox exploring Peter Coyote's yard.

wildlife as simply being a good neighbor. "Habitat for wildlife is continually shrinking—I can at least provide a way station," he says.

The animals have definitely noticed the welcome mat he has extended. His garden is simply the native landscape enhanced; it retains the memory of days when Roosevelt elk and grizzly bears freely roamed the area. Native wildlife—albeit smaller than the historical megafauna—still flock to his mini backyard nature reserve. Peter also supplements the native plants with birdfeeders. The well-stocked stash of sunflower seeds entice the titmice and juncos to visit, while goldfinches feed on his offerings of gourmet thistle.

Small mammals also make frequent appearances. He's witnessed raccoon and skunk families on parade in his yard (sometimes at the same time!), and one raccoon, named Monica, has raised her young in his garden for four years. A gray fox has become a regular resident—he once watched her, along with her three pups, drink from a clay water bowl on his deck.

After spending an afternoon with Peter at his home (nicknamed "the Tree House") it's obvious that he "walks the walk" of being a caretaker for wild things. The words of his friend the poet Gary Snyder perhaps best describe his philosophy: "Nature is not a place to visit. It is home."

Indeed, the natural world and his official dwelling seem indistinguishable, an extension of each other. From the road a series of winding staircases suspended among the redwood trees overlook ferns and other lush foliage in the creek bed below. Inside the home, you feel as if you were in the comforting embrace of a giant tree trunk. Peter describes the intent of the design: "My house and my garden are built as part of nature, not over it."

NORTHROP RANCH

During my first tour of her cattle ranch, Lynn Northrop drove me around her three thousand acres of rolling hills and oak woodlands in the Sierra foothills in an old Jeep, deftly shifting the long clutch and navigating steep inclines. As we rounded a corner on a dirt road, she suddenly hit the brakes. She jumped out and picked up a small western pond turtle that was trying to cross the road.

She placed the turtle safely on the other side. "These poor guys are having a hard time with the drought," she said. "They need a little extra help."

Lynn and her husband, Wayne, bought the ranch almost thirty years ago, mainly as a way to make a living in a rural landscape they both love. Raising cattle was a secondary goal, a means to an

Rancher Lynn Northrop helping a western pond turtle cross the road.

end of being able to preserve open space by operating a working ranch. "We were trying to decide how to work the land so we could make it our home," she says. "Farming was never going to be an option, as it's very slow to me, watching things grow. I am too hyper for that."

Lynn is an overachiever. Along with operating a ranch, both she and her husband have starred in several soap opera series for more than thirty years—Lynn is most well known for her character Lucy on *General Hospital.* In her spare time, she founded, built, curated, and still regularly staffs the Raymond Museum, which celebrates the history of the small community in which she lives. She's also a champion for sustainable and wildlife-friendly ranching. One year she invited me to attend the annual conference of the California Rangeland Conservation Coalition, and I was impressed with the sincere desire of those I met there to coexist with wildlife on their lands.

Lynn practices what she preaches on her ranch. They avoid overgrazing by rotating the cattle to manage access, don't use traps or poison baits, and recently fenced off forty acres as completely off-limits to grazing. As a result, three beavers appeared. "Everything that should be in the Sierra foothills we have on our ranch—mountain lions, coyotes, bobcats," she says. "That's what I want to be here. I love taking my binoculars and sitting and watching baby quail in spring when they first hatch, or a red-tailed hawk tussling with a snake. Last week we saw a dipper fly behind a waterfall. I love seeing wildlife just being themselves, and I like providing a place where they can simply live their lives."

DIANE ELLIS

Living in a city and with a small backyard, Diane Ellis didn't let herself get discouraged in her quest to garden for wildlife. As she observes, "A little really goes a long way. I don't have a lot of outdoor space, but you really don't need much space to help birds, bees, or butterflies. I get creative with pots and planters."

Diane has always been interested in wildlife and animals—wolves, marine mammals, and rescue dogs are just a few of her passions—but it wasn't until she took a conservation trip with the National Wildlife Federation to Belize that she realized she could make a difference for wildlife right at her own home in Sunset Beach. One of the NWF staff members on the trip told her about the Certified Wildlife Habitat designation, and she became intrigued with the idea that she could transform her small yard in a city into a mini wildlife reserve.

After visiting the local native plant nursery, she brought home some milkweed and, sure enough, the monarchs soon discovered her new habitat. "At first I thought the neighbors would think

Part of Diane Ellis's annual backyard monarch brood.

I was silly. But then they got interested and involved and started texting each other when the monarchs hatched. Some of them planted milkweed in their own yards," she says. As she expanded beyond those first few plants, redoing much of her yard space to be pollinator friendly, her monarch numbers have grown; last year she observed more than one hundred monarchs over the course of a summer making her garden their home. "It just gives me so much joy to see them flying around," she says, "and it's also important to life. And it's not hard to do. I always tell people: plant milkweed."

THE CITY OF CHULA VISTA

One of my first duties as the new California Director for the National Wildlife Federation in 2011 was to present to the City of Chula Vista its official certification for achieving the status of Community Wildlife Habitat.

I needed no better reminder of why I took the role with the NWF than the passion and enthusiasm of the dozens of people who attended the celebration at the city council meeting when Chula Vista became the fifty-third such community in the country and the third in the state of California. The city of 243,000 residents in San Diego County is also the second largest to become certified in the nation, behind only Austin in population.

Overall, an impressive 303 homes, 10 schools, 2 nature centers, 5 parks, 4 businesses, a botanical garden, a recreation center, the civic center, the grounds of a condominium complex, and the municipal golf course all achieved Certified Habitat status in Chula Vista.

Rice Canyon Open Space Preserve in Chula Vista.

After I presented awards in front of Mayor Cheryl Cox and the city council, Michelle Castagnola, who led the certification team for Chula Vista, organized a small party outside City Hall. I enjoyed talking with members of several key community and government groups, including the Chula Vista Garden Club, the Sweetwater Authority and Otay Water districts, and the Eastlake HOA—all of whom helped with the certification process. Pat Alfaro, one of the community members who worked on the East-lake HOA project, attended the ceremony. "I get such a feeling of accomplishment from helping create a habitat for birds, bees, butterflies, and other small creatures," she said. "We did a great thing for ourselves, our community, and the local wildlife."

FEBRUARY-MARCH 2016

NATIONAL
WILDLIFE®

NWF CELEBRATES
80
YEARS

Cover of the anniversary issue of *National Wildlife* magazine, celebrating eighty years of the National Wildlife Federation.

Get Involved

THE NATIONAL WILDLIFE FEDERATION

One of the oldest and largest conservation groups in the country, the National Wildlife Federation is a strong voice for wildlife, dedicated to protecting wildlife and habitat and inspiring young people today to become conservation-minded adults. Visit www.nwf.org for more information about the impact of the NWF's work and how you can help wildlife by joining our more than six million supporters nationwide. Take action by:

Becoming a member. Your membership will help protect at-risk wildlife and the wild places they depend upon for survival. Membership includes a free subscription to *National Wildlife* magazine.

Gardening for Wildlife. Join millions of novice, hobby, and master gardeners across the country in attracting wildlife to your backyard, schoolyard, business, place of worship, or even your city, and then register as an official National Wildlife Federation Certified Wildlife Habitat®.

Subscribing to our award-winning children's magazines *Ranger Rick*® **and** *Ranger Rick, Jr.*® These publications are intended to instill a lifelong passion for nature and a love of reading, and to promote getting outdoors for great activities.

Joining other young adults in the NWF EcoLeaders Initiative. This online community offers college-level students the space to create, to share, and to be recognized for their leadership efforts and for sustainability projects and campaigns. Learn more at www.nwfecoleaders.org.

Enrolling your school in our free and critically acclaimed Eco-Schools USA program. You can green your campus and your curriculum through this student-led initiative. Learn more at www.nwf.org/Eco-SchoolsUSA.

THE NATIONAL WILDLIFE FEDERATION IN CALIFORNIA

California and its diverse communities play a critical role in building a stronger conservation movement. The NWF leads a number of important programs to help protect the state's remarkable wildlife, and our initiatives are designed to help both the wildlife and people of California thrive. Some of our projects include:

#SaveLACougars. We're spearheading efforts to build what could be the world's largest wildlife crossing, to help save a population of mountain lions with our partner, the Santa Monica Mountains Fund, and a large coalition and community of organizations, businesses, individuals, and elected officials. For more information, visit www.savelacougars.org.

The Return of the Porpoise to San Francisco Bay. We're helping to support the work of Golden Gate Cetacean Research's study of porpoises, dolphins, and other marine mammals, including how the changing waters impact their survival. For more information, visit www.sfbayporpoises.org.

Protecting Silicon Valley's Wildside. Gray foxes and other wildlife need corridors to survive. Together with our partner, the Urban Wildlife Research Project, we want to implement a San Francisco Bay Wildlife Connection Corridor using high-tech campuses and other private lands. For more information, visit www.svwildlife.org.

For additional information on the NWF's conservation work in California, visit www.nwf.org/california, and follow us on Facebook and Twitter at NWFCalifornia.

WILDLIFE AGENCIES AND PROFESSIONALS

Wildlife in California (and around the globe) depend on various federal, state, and local agencies and research institutions, many of which are named in this book. Healthy wildlife populations rely on the strong and visionary management of these organizations, as well as adequate funding to make their ideas a reality. When you vote, be sure to support those programs and political candidates who represent the best interests of wildlife.

The California Department of Fish and Wildlife manages the state's diverse fish, wildlife, and plant resources, and the habitats upon which they depend. It works with a multitude of public and private partners to carry out its mission and offers an array of public programs and educational resources. For more information, visit https://www.wildlife.ca.gov/.

WILDLIFE AND WILDLIFE RESCUE NONPROFIT ORGANIZATIONS

Nonprofit partners working on conservation and wildlife rescue—including many featured in this book—are also vital to the future of wildlife. Although it's outside the scope of this publication to list the thousands of organizations doing great work in California, please be sure to support their efforts and make a difference for wildlife by donating money and/or volunteering your time.

ACKNOWLEDGMENTS

Over some fried plantains and pupusas at a Berkeley Salvadoran restaurant we often frequent, I related the story of P-22 to Malcolm Margolin, founder of Heyday, telling him how the plight of this lonely cougar rallied an entire city. And as I shared other stories of wildlife becoming our "neighbors" and how I thought these stories, taken in the context of accumulation, attested to a change in our relationships with wildlife, this book was born. Malcolm is a trusted friend and mentor, and his guidance has led me down many foolish and not-so-foolish paths for almost two decades; I cannot thank him enough for both instigating and supporting this endeavor.

Any book is a marathon, and without my amazing support team, *When Mountain Lions Are Neighbors* never would have been completed. Along with Malcolm, all the folks at Heyday put their hearts and souls into every book they produce. Gayle Wattawa forgave me my popular-culture addiction, and her immense editing talent combined with the fine eye of copyeditor Lisa K. Marietta propelled this book to another level. Diane Lee, Lillian Fleer, and Patricia Wakida especially held my hand through this process, and Sylvia Linsteadt and Michael Drake did some early research. But all the folks at Heyday deserve thanks—it's a community that produces amazing things when working together.

Many wonderful people at my organization, the National Wildlife Federation, also offered their assistance to help make this book a reality. I can't thank the NWF's president and CEO Collin O'Mara enough for supporting the project and also writing the foreword, and Amanda McKnight, Andy Buchsbaum, Dirk Sellers, and Molly Judge were instrumental in moving this forward. My trusty partner in crime, Leigh Wyman, has been with me every step of the way, did amazing research for this book, and had great insights into early drafts of the work. David Mizejewski, Doug Inkley, Jim Lyon, Laura Hickey, Maureen Smith, Mary Phillips, and Ben Kota provided valuable reviews. Current and former NWF staff Carolyn Millard, Hilary Falk, Jane Kirchner, Les Welsh, Anne Bolen, Debbie Anderson, Michael Morris, Michael Cooper, Claire Megginson, Jill King, Jessie Yuhaniak, Jennifer Janssen, Becky McIntire, Dani Tinker, Danielle Brigida, John Kostyack, Tori Leach, Cindy Golos, Jenni Lopez, Brandon Miller, and Barbara McIntosh, along with NWF board member Julia Reed Zaic, also deserve kudos. I owe a special note of thanks to my former boss, Tony Caligiuri, then the NWF's senior vice president of conservation, who after reading the first chapter of the book really championed this project.

A huge thanks to all the scientists, agency staff, wildlife enthusiasts, and others who made time to take me out in the field or do interviews—I am indebted to them all: Mark Abramson, Pat Alfaro, Kenneth Balcomb, Erik Beever, Patricia Betancourt, Bruce Bonafede, Erin Boydston, Joseph Brandt, Christy Brigham, Daniel Byers, Kevin Casbeer, Michelle Castagnola, Don Ciota, Clif Clark, Ron Clausen, Jane Coloccia, Dan Cooper, Peter Coyote, Michelle Dennehy, Eddie Dunbar, Art Eck, Diane Ellis, Allen Fish, Camilla Fox, Mark Gallagher, Scott Gediman, Jason Goldman, Dan and Susan Gottlieb, Rob Grasso, Noah Greenwald, Margot Griswold, Sara Leon Guerrero, Gerry Hans and Mary Button, Jaymi Heimbuch, Keli Hendricks, Brian Henen, Lila Higgins, Jack Hopkins, Erin Hunt, Megan Isadore, Dan Jensen, Kevin Joe, Bill Keener, Greg Kerekez, Rob Klinger, Kerry Kriger, Kate Kuykendall, Benjamin Landis, Richard Lanman, Ryan Leahy, Bill Leikam, Anne Lombardo, Travis Longcore, Rue Mapp, Faith Margaret, the staff of the Marine Mammal Center, Rachel Mazur, Linda Mazzu, John McCammon, Kate McCurdy, Zara McDonald, Michelle McGurk, Paul McKim, Tom Medema, Joe Medley, Ally Nauer, Sharon Negri, Don Neubacher, Craig Newmark, Lynn Northrop,

Miguel Ordeñana, Greg Pauly, Heidi Perryman, Greg Randall, Seth Riley, Tora Rocha, Caitlin Lee-Roney, Jacqueline Rooney, Bradley Rumble, Virginia M. Sanchez, Cindy Sandoval, Elizabeth Sarmiento, Clemens Schenk, Laurel Serieys, Jeff Sikich, Alison Simard, Lili Singer, Alexis Smith, Douglas Smith, Diane Shader Smith, Michael Starkey, Margie Steigerwald, Jonathan Stern, Glenn Stewart, Sarah Stock, Izzy Szczepaniak, David Szymanski, Craig Thompson, Steve Thompson, Jordan Traverso, Jeffrey Trust, Karin Vardaman, Dean Waziry, Marc Webber, Amaroq Weiss, Ken White, Chris Wilmers, and David Wyatt.

Kathryn Bowers was the first person to read the story of P-22 and encouraged me to continue; I cannot thank her enough for rescuing me from that common despair that writers get when starting a work—the "this sucks" syndrome. John Griffith also provided invaluable editing on the first draft, and Jon Christensen, Madelyn and Jerry Jackrel, Dan Gerber, Mary Ellen Hannibal, Michelle Hansen, Teresa Cirolia, and Matteo Fiori also were generous enough to read early drafts and provide me with the confidence to continue. And my thanks to all of my Facebook friends, who helped build this book for years by enthusiastically commenting on all of the stories I shared.

I cannot express enough appreciation to all the photographers who provided images for this book (see photo credits on page 211). All of the images were amazing, and I wish we could have used every one that was submitted.

Photographer Steve Winter is my hero, and his willingness to donate his images of P-22 for this book and our #SaveLACougars campaign shows his immense generosity. As soon as I saw P-22 with the lights of LA in the background, I knew I had my cover photo.

A special note of thanks and appreciation goes to the late photographer Rebecca Jackrel, who donated images for the work but also offered something beyond that: inspiration. In my all-too-short friendship with her, I came to admire her tenacity in advocating for conservation in her work, from publishing her book *The Ethopian Wolf,* to volunteering at the Marine Mammal Center with her husband, Lee, to visiting all seven continents to help raise awareness for a variety of imperiled wildlife. I admired her passion and courage. I miss her greatly.

Friends—there are just too many of you to name, but you all helped in some way, from letting me crash on your couch, to cooking me dinner, to emailing me inspiring wildlife stories, to accompanying me on an adventure to investigate a story.

And last but not least, the most important people that helped in bringing this book to fruition is my family. My parents, Bill and Martha, instilled in me a lifetime love of wildlife that eventually gave birth to this book. And the house- and dog-sitting they provided, along with my brother, Kevin, allowed me to rove around California for research.

Finally, this manuscript would not have been completed without my husband and best friend, Mike Bergstrom, who enabled me to write this book—in my "spare time" on top of an already full-time job—with his gift of support. He took over the household duties of cooking, cleaning, doing laundry (and I haven't yet picked up the slack even though the book is finished), tended to the needs of our four dogs and two cats, and provided much-appreciated endless shoulder massages after my long days at the computer. He also drank the Kool-Aid in the process and has become a full-fledged naturalist; he's turned out to be a pretty good pika spotter, he can use the word "lagomorph" confidently in a sentence, and he now shares my delight when the tadpoles arrive every year in our frog pond.

PHOTO CREDITS

Page v: Miguel Ordeñana
Page xix: Rebecca Abbey
Pages xxvi, 28: Steve Winter
Page 5: Courtesy of NASA/JPL-
Caltech
Pages 10, 16, 25, 29, 31, 82, 88, 91, 94,
97, 102, 176: National Park Service
Page 22: Clark Stevens/Raymond
Garcia, Resource Conservation
District of the Santa Monica
Mountains
Page 30: Christopher Stills
Pages 32, 75, 95: Johanna Turner
Page 33: LionsandTigers.org
Page 34: Lila Higgins
Page 35: Official USMC photo by Kelly
O'Sullivan
Pages 36, 47, 48,: Bill Keener
Page 42: © Solvin Zankl/Visuals
Unlimited/Corbis
Page 45: David Liittschwager
Pages 51, 58, 62: Bill Keener/Golden
Gate Cetacean Research, acquired
under NOAA Fisheries permit
LOC#15477
Pages 63, 66, 70: Rebecca Jackrel
Page 64: Arwen Kreber-Mapp
Page 65: Cheryl Reynolds/Worth A
Dam/Martinezbeavers.org
Page 67: George Eade
Page 68: Marine Mammal Center
Page 69: Jaymi Heimbuch/The Natural
History of the Urban Coyote
Page 85: Robert E. Riggins
Page 100: Phil Frank/www.
farleycomicstrip.com
Page 101: Laura Cunningham
Page 103: Steven M. Bumgardner
Page 104: Morgan Heim
Pages 105, 107, 140, 141, 165, 173, 190,
193, 197: Beth Pratt-Bergstrom
Page 106: Joe Medley
Pages 108, 120, 133: Greg Kerekez
Page 111: Jeffrey Ferland

Page 112: Karl Frankowski
Page 115: Mark Zuckerberg; original
photo by Josh Frankel
Page 118: Richard Zadorozny
Pages 125, 126: William C. Leikam—
The Fox Guy
Page 135: Photo courtesy of CSU–
Stanislaus Endangered Species
Recovery Program. Photo by
Francesca Ferrara.
Page 136: Nick Dunlop
Page 137, 175: Jerry Ting
Page 138: With permission of the
Peninsula Humane Society and
SPCA
Page 139: Rollin Coville
Page 142: John Gabois
Pages 147, 154, 156: US Fish and
Wildlife
Page 148: Oregon Department of
Fish and Wildlife, and California
Department of Fish and Wildlife
Page 157: California Department of
Fish and Wildlife
Page 169: California Wolf Center
Page 170: © Clemens Schenk, all rights
reserved
Page 171: Amanda Schuffleburger,
Biologist—Sierra Pacific Industries,
Meso-carnivore survey camera
Page 172, 189: Courtesy of California
Conservation Corps
Page 174: David T. Wyatt
Page 180: Elizabeth Young
Page 182: Eric Olson
Pages 184, 202: National Wildlife
Federation
Page 186: Left side, Jennifer MaHarry.
Right side, Susan Gottlieb.
Page 195: Peter Coyote
Page 199: Diane Ellis
Page 201: Photo courtesy of the city
of Chula Vista
Page 212: Bill Pratt

ABOUT THE AUTHOR

Beth Pratt-Bergstrom has worked in environmental leadership roles for more than twenty-five years, and in two of the country's largest national parks: Yosemite and Yellowstone. As the California Director for the National Wildlife Federation, she says, "I have the best job in the world—advocating for the state's remarkable wildlife." Her conservation work has been featured by the *New Yorker,* the *Wall Street Journal, BBC World Service, CBS This Morning,* the *Los Angeles Times,* and NPR, and she has written for CNN.com, *Boom: A Journal of California, Yellowstone Discovery, Yosemite Journal, Darling,* and *Inspiring Generations: 150 Years, 150 Stories in Yosemite.* She is the author of the novel *The Idea of Forever* and the official *Junior Ranger Handbook* for Yosemite. Beth lives outside of Yosemite, "my north star," with her husband, four dogs, two cats, and the mountain lions, bears, foxes, frogs, and other wildlife that frequent her NWF Certified Wildlife Habitat backyard.

Find Beth online at
www.bethpratt.com
Facebook: bethpratt1
Twitter: @bethpratt
Instagram: yosemitebethy

The author with an elephant seal at Año Nuevo State Park.

HEYDAY
into California

About Heyday

Heyday is an independent, nonprofit publisher and unique cultural institution. We promote widespread awareness and celebration of California's many cultures, landscapes, and boundary-breaking ideas. Through our well-crafted books, public events, and innovative outreach programs we are building a vibrant community of readers, writers, and thinkers.

Thank You

It takes the collective effort of many to create a thriving literary culture. We are thankful to all the thoughtful people we have the privilege to engage with. Cheers to our writers, artists, editors, storytellers, designers, printers, bookstores, critics, cultural organizations, readers, and book lovers everywhere!

We are especially grateful for the generous funding we've received for our publications and programs during the past year from foundations and hundreds of individual donors. Major supporters include:

Advocates for Indigenous California Language Survival; Anonymous (3); Judith and Phillip Auth; Carrie Avery and Jon Tigar; Judy Avery; Dr. Carol Baird and Alan Harper; Paul Bancroft III; Richard and Rickie Ann Baum; BayTree Fund; S. D. Bechtel, Jr. Foundation; Jean and Fred Berensmeier; Joan Berman and Philip Gerstner; Nancy Bertelsen; Barbara Boucke; Beatrice Bowles; Jamie and Philip Bowles; John Briscoe; David Brower Center; Lewis and Sheana Butler; Helen Cagampang; California Historical Society; California Rice Commission; California State Parks Foundation; California Wildlife Foundation/California Oaks; The Campbell Foundation; Joanne Campbell; Candelaria Fund; John and Nancy Cassidy Family Foundation; James and Margaret Chapin; Graham Chisholm; The Christensen Fund; Jon Christensen; Cynthia Clarke; Lawrence Crooks; Community Futures Collective; Lauren and Alan Dachs; Nik Dehejia; Topher Delaney; Chris Desser and Kirk Marckwald; Lokelani Devone and Annette Brand; J.K. Dineen; Frances Dinkelspiel and Gary Wayne; The Roy & Patricia Disney Family Foundation; Tim Disney; Doune Trust; The Durfee Foundation; Michael Eaton and Charity Kenyon; Endangered Habitats League; Marilee Enge and George Frost; Richard and Gretchen Evans; Megan Fletcher; Friends of the Roseville Public Library; Furthur Foundation; John Gage and Linda Schacht; Wallace Alexander Gerbode Foundation; Patrick Golden; Dr. Erica and Barry Goode; Wanda Lee Graves and Stephen Duscha; Walter & Elise

Haas Fund; Coke and James Hallowell; Theresa Harlan; Cindy Heitzman; Carla Hills and Frank LaPena; Leanne Hinton and Gary Scott; Charles and Sandra Hobson; Nettie Hoge; Donna Ewald Huggins; Inlandia Institute; JiJi Foundation; Claudia Jurmain; Kalliopeia Foundation; Marty and Pamela Krasney; Guy Lampard and Suzanne Badenhoop; Thomas Lockard and Alix Marduel; David Loeb; Thomas J. Long Foundation; Judith Lowry-Croul and Brad Croul; Bryce and Jill Lundberg; Sam and Alfreda Maloof Foundation for Arts & Crafts; Manzanar History Association; Michael McCone; Nion McEvoy and Leslie Berriman; The Giles W. and Elise G. Mead Foundation; Moore Family Foundation; Michael Moratto and Kathleen Boone; Seeley W. Mudd Foundation; Karen and Thomas Mulvaney; Richard Nagler; National Wildlife Federation; Native Arts and Cultures Foundation; The Nature Conservancy; Nightingale Family Foundation; Steven Nightingale and Lucy Blake; Northern California Water Association; Panta Rhea Foundation; Julie and Will Parish; Ronald Parker; Pease Family Fund; Jean and Gary Pokorny; Jeannene Przyblyski; James and Caren Quay, in honor of Jim Houston; Steven Rasmussen and Felicia Woytak; Susan Raynes; Robin Ridder; Spreck Rosekrans and Isabella Salaverry; Alan Rosenus; The San Francisco Foundation; San Francisco Architectural Heritage; Toby and Sheila Schwartzburg; Mary Selkirk and Lee Ballance; Ron Shoop; The Stephen M. Silberstein Foundation; Ernest and June Siva; Stanley Smith Horticultural Trust; William Somerville; Carla Soracco and Donna Fong, in honor of Barbara Boucke; Radha Stern and Gary Maxworthy; Liz Sutherland; Roselyne Swig; Thendara Foundation; TomKat Charitable Trust; Jerry Tone and Martha Wyckoff; Sonia Torres; Michael and Shirley Traynor; The Roger J. and Madeleine Traynor Foundation; Lisa Van Cleef and Mark Gunson; Stevens Van Strum; Patricia Wakida; Marion Weber; Sylvia Wen and Mathew London; Valerie Whitworth and Michael Barbour; Cole Wilbur; Peter Wiley and Valerie Barth; The Dean Witter Foundation; and Yocha Dehe Wintun Nation.

Getting Involved

To learn more about our publications, events and other ways you can participate, please visit www.heydaybooks.com.